Galois Theory

CHAPMAN AND HALL
MATHEMATICS SERIES

Edited by Professor R. Brown.
Head of the Department of Pure Mathematics,
University College of North Wales, Bangor,
and Dr Michael Dempster,
Lecturer in Industrial Mathematics
and Fellow of Balliol College, Oxford

A Preliminary Course in Analysis
R. M. F. Moss *and* G. T. Roberts

Elementary Differential Equations
R. L. E. Schwarzenberger

A First Course on Complex Functions
G. J. O. Jameson

Rings, Modules and Linear Algebra
B. Hartley *and* T. O. Hawkes

Regular Algebra and Finite Machines
J. H. Conway

Complex Numbers
W. H. Cockcroft

Topology and Normed Spaces
C.J.O. Jameson

Introduction to Optimization Methods
P.R. Adby *and* M.A.H. Dempster

Galois Theory

IAN STEWART

Lecturer, Mathematics Institute,
University of Warwick, Coventry

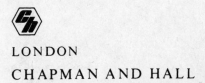

LONDON

CHAPMAN AND HALL

First published 1973
by Chapman and Hall Ltd
11 New Fetter Lane, London EC4P 4EE
Reprinted 1975 and 1976

© *1973 Ian Stewart*

Filmset by Keyspools Ltd, Golborne, Lancs
Printed in Great Britain by
Redwood Burn Limited
Trowbridge & Esher

ISBN 0 412 10800 3

Published in the U.S.A.
by Halsted Press, a Division
of John Wiley & Sons, Inc.
New York

Preface

Galois Theory is a showpiece of mathematical unification, bringing together several different branches of the subject and creating a powerful machine for the study of problems of considerable historical and mathematical importance. This book is an attempt to present the theory in such a light, and in a manner suitable for second and third year under-graduates.

The central theme is the application of the Galois group to the quintic equation. As well as the traditional approach by way of the 'general' polynomial equation I have included a direct approach which demonstrates the insolubility by radicals of a specific quintic polynomial with integer coefficients, which I feel is a more convincing result. The abstract Galois theory is set in the context of arbitrary field extensions, rather than just subfields of the complex numbers; the resulting gain in generality more than compensates for the extra work required. Other topics covered are the problems of duplicating the cube, trisecting the angle, and squaring the circle; the construction of regular polygons; the solution of cubic and quartic equations; the structure of finite fields; and the 'fundamental theorem of algebra'. The last is proved by almost purely algebraic methods, and provides an interesting application of Sylow theory.

In order to make the treatment as self-contained as possible, and to bring together all the relevant material in a single volume, I have included several digressions. The most important of these is a proof of the transcendence of π, which every mathematician should see at least once in his life. There is a discussion of Fermat numbers, to emphasize that the problem of regular polygons, although reduced to a simple-looking question in number theory, is by no means completely solved. A construction for the regular 17-gon is given, on the grounds that such an unintuitive result requires more than just an existence proof.

Much of the motivation for the subject is historical, and I have taken the opportunity to weave historical comments into the body of the book where appropriate. There are two sections of purely historical matter: a short sketch of the history of polynomials, and a biography of Évariste Galois. The latter is culled from several sources (listed in the references) of which by far the most useful and accurate is that of Dupuy [37].

I have tried to give plenty of examples in the text to illustrate the general theory, and have devoted one chapter to a detailed study of the Galois group of a particular field extension. There are nearly 200 exercises, with 20 harder ones for the more advanced student.

Many people have helped, advised, or otherwise influenced me in writing this book, and I am suitably grateful to them. In particular my thanks are due to Rolph Schwarzenberger and David Tall, who read successive drafts of the manuscript; to G. A. Paxson and Professor R. M. Robinson for information incorporated in the table of Fermat numbers; to Mr L. Bulmer and the staff of the University of Warwick Library for locating documents relevant to the historical aspects of the subject; to Professor Ronald Brown for editorial guidance and much good advice; and to the referee who pointed out a multitude of sins of omission and commission on my part, whose name I fear will forever remain a mystery to me, owing to the

system of secrecy without which referees would be in
continual danger of violent retribution from indignant
authors.

University of Warwick,   IAN STEWART
Coventry,
April, 1972.

Notes to the reader

Theorems, lemmas, propositions, corollaries, and the like are numbered consecutively within chapters by numbers of the form **m.n** where **m** is the chapter number and **n** indicates the position within the chapter.

Exercises are given at the end of each chapter (with two exceptions) and are numbered in a similar fashion. There are also 20 harder exercises in a separate section at the end. Solutions are given to some of the exercises, mostly those whose solution can be made brief.

Definitions are usually, but *not* always, signalled by the word **Definition**.

Equations which need to be referred to are numbered (n) at the right-hand side of the page, the numbering starting afresh with each chapter.

References are given at the back, and are signalled in the text by numbers of the form $[m]$.

Structure

Each brick represents a chapter. Mathematical dependence of chapters corresponds to structural dependence of bricks.

For a short course aimed directly at the insolubility of the quintic equation the sequence of Chapters **1 2 3 4 7 8 9**

10 11 13 14 is recommended. Alternatively the third sub-section of **13** may be omitted, together with the second half of **14**, and Chapter **15** substituted.

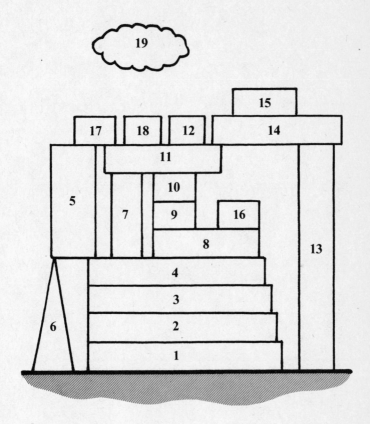

Contents

Historical
introduction

Polynomial equations have a lengthy history. A Babylonian tablet of c. 1600 B.C. poses problems which reduce to the solution of quadratic equations (Midonick [44] p. 48); and it is clear from the tablets that the Babylonians possessed methods of solving them (Bourbaki [32] p. 92) although they had no algebraic notation with which to express their solution. The ancient Greeks solved quadratics by geometrical constructions, but there is no sign of an algebraic formulation until at least 100 A.D. (Bourbaki [32] p. 92). They also had methods applicable to cubic equations, involving points of intersection of conics. Algebraic solutions of the cubic were unknown, and in 1494 Pacioli ended his *Summa di Arithmetica* with the remark that the solution of the equations $x^3 + mx = n$ and $x^3 + n = mx$ was as impossible at the existing state of knowledge as squaring the circle.

The Renaissance mathematicians at Bologna discovered that the solution of the cubic could be reduced to that of three basic types: $x^3 + px = q$, $x^3 = px + q$, $x^3 + q = px$. They were forced to distinguish these cases because they did not recognize the existence of negative numbers. Scipio del Ferro is believed on good authority (Bortolotti [30]) to have solved all three types; he certainly passed on his method for one type to a student, Fior. News of the solution leaked out, and others were encouraged to try their hand; and solutions were rediscovered by Niccolo Fontana (nicknamed Tartaglia) in 1535. Fontana demonstrated his methods in a public competition with Fior, but refused to reveal the details. Finally he was persuaded to tell them to the physician Girolamo Cardano, having first sworn him to secrecy. But when Cardano's *Ars Magna* appeared in 1545 it contained a complete discussion of Fontana's solution – with full acknowledgement to the discoverer. Although Cardano claimed motives of the highest order [33] Fontana was justifiably annoyed, and in the ensuing wrangle the history of the discovery became public knowledge.

The *Ars Magna* also contained a method, due to Ludovico Ferrari, of solving the quartic equation by reducing it to a cubic.

All the formulae discovered had one striking property, which can be illustrated by Fontana's solution of $x^3 + px = q$:

$$x = \sqrt[3]{\frac{q}{2} + \sqrt{\frac{p^3}{27} + \frac{q^2}{4}}} + \sqrt[3]{\frac{q}{2} - \sqrt{\frac{p^3}{27} + \frac{q^2}{4}}}$$

The expression is built up from the coefficients by repeated addition, subtraction, multiplication, division, and extraction of roots. Such expressions became known as *radical* expressions. Since all equations of degree ≤ 4 were now solved, it was natural to ask how the quintic equation could be solved by radicals.

Many mathematicians attacked the problem. Tschirnhaus claimed a solution, recognized as fallacious by Leibniz. Euler failed to solve the problem but found new methods for the quartic. Lagrange took an important step in 1770 when he unified the separate tricks used for the equations of degree ≤ 4. He showed that they depended on finding functions of the roots of the equation which were unchanged by certain permutations of those roots; and he showed that this approach *failed* when tried on the quintic.

A general feeling that the quintic could not be solved by radicals was now in the air; and in 1813 Ruffini attempted to give a proof of the impossibility. His paper appeared in an obscure journal, with several gaps in the proof (Bourbaki [32] p. 103) and attracted little attention. The question was finally settled by Abel in 1824, who proved conclusively that the general quintic equation was insoluble by radicals.

The problem now arose of finding a way of deciding whether or not a given equation could be solved by radicals. Abel was working on it when he died in 1829. In 1832 a young Frenchman, Évariste Galois, was killed in a duel. He had for some time sought recognition for his mathematical theories, submitting three memoirs to the Academy of Sciences in Paris. They were all rejected; and his work appeared to be lost to the mathematical world. Then, on 4 July 1843, Joseph Liouville addressed the academy. He opened with these words:

'*I hope to interest the Academy in announcing that among the papers of Évariste Galois I have found a solution, as precise as it is profound, of this beautiful problem: whether or not it is soluble by radicals. . . .*'

The life of Galois

Évariste Galois was born at Bourg-la-Reine near Paris on 25 October 1811. His father Nicolas-Gabriel Galois was a Republican [43] and head of the village liberal party; after the return to the throne of Louis XVIII in 1814 he became mayor. Évariste's mother Adelaide-Marie (née Demante) was the daughter of a jurisconsult. She was a fluent reader of Latin, thanks to a solid education in religion and the classics.

For the first twelve years of his life Galois was educated by his mother, who passed on to him a thorough grounding in the classics. His childhood appears to have been a happy one [43]. At the age of 10 he was offered a place at the college of Reims, but his mother preferred to keep him at home. In October 1823 he entered the lycée Louis-le-Grand. During his first term there the students rebelled and refused to chant in chapel, and 100 of them were expelled [43].

Galois performed well during his first two years at school, obtaining first prize in Latin; but then boredom set in. He was made to repeat the next year's classes, but this simply aggravated the tedium. It was during this period that Galois began to take a serious interest in mathematics. He came across a copy of Legendre's *Éléments de Géométrie*, a classic text which broke with the Euclidean tradition of school geometry. It is said [28] that he read it 'like a novel' and mastered it in one reading. The school algebra texts could not compete with Legendre's masterpeiece, and Galois turned instead to the original memoirs of Lagrange and Abel. At the age of 15 he was reading material written for professional mathematicians. But his classwork remained uninspired; it would seem that he had lost all interest in it. His teachers misunderstood him and accused him of *affecting* ambition and originality.

Galois was an untidy worker, as can be seen from some of his manuscripts [31]; and he tended to work in his head, committing only the results of his deliberations to paper. His teacher Vernier begged him to work systematically, but Galois ignored his advice. Without adequate preparation he took the competitive examination for entrance to the École Polytechnique. A pass would have ensured his success, for the Polytechnique was the breeding-ground of French mathematics. He failed. Two decades later Terquem (editor of the *Nouvelles Annales des Mathématiques*) advanced the following explanation: 'A candidate of superior intelligence is lost with an examiner of inferior intelligence. Because they do not understand me, *I* am a barbarian. . . .'

In 1828 Galois entered the École Normale (a pale shadow of the Polytechnique) and attended an advanced class in mathematics under Richard, who was very sympathetic towards him. Richard was of the opinion that Galois should be admitted to the Polytechnique without examination [28]. The following year saw the publication of Galois's first paper, on continued fractions; though competent it held no hint of genius (see [38]). Meanwhile Galois had been making fundamental discoveries in the theory of poly-nomial equations, and he submitted some of his results to the Academy of Sciences. The referee was Cauchy, who had already published work on the behaviour of functions under permutation of the variables, a central theme in Galois's theory. Cauchy rejected the memoir, and another presented eight days later fared the same. The manuscripts were lost, never to be seen again [43].

The same year held two more disasters. On 2 July 1829 Galois's father committed suicide after a bitter political dispute with the village priest. A few days later Galois sat again for entrance to the Polytechnique – his final chance. There is a legend [28, 37] that he lost his temper and threw an eraser into the examiner's face. But according to Bertrand [29] this tradition is false. The examiner Dinet asked Galois to outline the theory of 'arithmetical logarithms' . . . and Galois informed him that there were no *arithmetical* logarithms. Dinet failed him.

In February 1830 Galois presented his researches to the Academy of Sciences in competition for the Grand Prize in Mathematics – the pinnacle of mathematical honour. The work has since been judged more than worthy of the prize [28]. The manuscript reached the secretary, Fourier, who took it home for perusal. But he died before reading it, and the manuscript could not be found among his papers. According to Dupuy [37] Galois considered that the repeated losses of his papers were not the effect of mere chance. He saw them as the effect of a society in which genius was condemned to an eternal denial of justice in favour of mediocrity; and he blamed the politically oppressive Bourbon regime.

Charles X had succeeded Louis XVIII in 1824. In 1827 the liberal opposition had made electoral gains; and in 1830 more elections were held, giving the oppositing a majority. Charles, faced with abdication, attempted a *coup d'état*. On 25 July he issued his notorious *Ordonnances* suppressing the freedom of the Press. The populace was in no mood to tolerate these steps, and revolted. The uprising lasted three days, after which as a compromise the Duke of Orléans, Louis-Philippe, was made king. During these three days, while the students of the Polytechnique were making history in the streets, Galois and his fellow students were locked in by Guignault, director of the École Normale. Galois was incensed and subsequently wrote a blistering attack [37] on him in the *Gazette des Écoles*, signing the letter with his full name. The editor removed the signature, and Galois was expelled as a result of his 'anonymous' letter [36]. (There is an interesting and detailed discussion of the circumstances in Dupuy [37].)

On 13 January 1831 Galois tried to set up as a private teacher of mathematics, offering a course in advanced algebra. He met with little success. On 17 January he sent once more a memoir to the Academy: *On the conditions of solubility of equations by radicals*. Cauchy was no longer in Paris, and Poisson and Lacroix were appointed referees. After two months Galois had heard no word from them, and he wrote to the President of the Academy asking what was happening. He received no reply.

He joined the artillery of the National Guard, a Republican organization. Soon afterwards its officers were arrested as conspirators, but acquitted by the jury. The artillery was disbanded by royal order. On 9 May a banquet was held in protest; the proceedings became more and more riotous, and Galois proposed a toast to Louis-Philippe with an open knife in his hand. His companions interpreted this as a threat on the king's life, applauded mightily, and ended up dancing and shouting in the street. The following day Galois was arrested. At the trial he admitted everything, but claimed that the toast proposed was actually 'To Louis-Philippe, *if he turns traitor*,' and that the uproar had drowned the last phrase. The jury acquitted him, and he was freed on 15 June.

On 4 July he heard the fate of his memoir. Poisson declared it 'incomprehensible'. The report (reprinted in full in [49]) ended as follows.

'We have made every effort to understand Galois's proof. His reasoning is not sufficiently clear, sufficiently developed, for us to judge its correctness, and we can give no idea of it in this report. The author announces that the proposition which is the special object of this memoir is part of a general theory susceptible of many applications. Perhaps it will transpire that the different parts of a theory are mutually clarifying, are easier to grasp together rather than in isolation. We would then suggest that the author

should publish the whole of his work in order to form a definitive opinion. But in the state which the part he has submitted to the Academy now is, we cannot propose to give it approval.'

On 14 July Galois was at the head of a Republican demonstration, wearing the uniform of the disbanded artillery, carrying a knife and a gun. He was arrested on the Pont-Neuf; convicted of illegally wearing a uniform [28]; and sentenced to six months' imprisonment in the jail of Sainte-Pélagie. He worked for a while on his mathematics; then in the cholera epidemic of 1832 he was transferred to a hospital. Soon he was put on parole.

Along with his freedom he experienced his first and only love affair, with one Mlle Stéphanie D. The surname is unknown; it appears in one of Galois's manuscripts, but heavily obliterated. There is much mystery surrounding this interlude, which has a crucial bearing on subsequent events. Fragments of letters [31] indicate that Galois was rejected and took it very badly. Not long afterwards he was challenged to a duel, ostensibly because of his relationship with the girl. Again the circumstances are veiled in mystery. One school of thought (Bell [28], Kollros [43]) asserts that the girl was used as an excuse to eliminate a political opponent on a trumped-up 'affair of honour'. In support of this we have the express statement of Alexandre Dumas (in his *Mémoires*) that one of the opponents was Pécheux D'Herbinville. But Dalmas [36] cites evidence from the police report suggesting that the other duellist was a Republican, possibly a revolutionary comrade of Galois's; and that the duel was exactly what it appeared to be. And this theory is largely borne out by Galois's own words on the matter [31]:

'I beg patriots and my friends not to reproach me for dying otherwise than for my country. I die the victim of an infamous coquette. It is in a miserable brawl that my life is extinguished. Oh! why die for so trivial a thing, for something so despicable! . . . Pardon for those who have killed me, they are of good faith.'

On the same day, 29 May, the eve of the duel, he wrote his famous letter to his friend Auguste Chevalier, outlining his discoveries; later published by Chevalier in the *Revue Encyclopédique*. In it he sketched the connection between groups and polynomial equations, stating that an equation is soluble by radicals provided its group is soluble. But he also mentioned many other ideas, about elliptic functions and the integration of algebraic functions; and other things too cryptic to be identifiable. It is in many ways a pathetic document, with scrawled comments in the margins: 'I have no time!'

The duel was with pistols at 25 paces. Galois was hit in the stomach, and died a day later on 31 May of peritonitis. He refused the office of a priest. On 2 June 1832 he was buried in the common ditch at the cemetery of Mont-parnasse.

His letter to Chevalier ended with these words:

'Ask Jacobi or Gauss publicly to give their opinion, not as to the truth, but as to the importance of these theorems. Later there will be, I hope, some people who will find it to their advantage to decipher all this mess. . . .'

Glossary
of Symbols

CHAPTER ONE

Background

The object of this introductory chapter is to provide a short account of the foundations upon which Galois Theory rests, namely: the concepts 'ring', 'field', 'integral domain', 'ideal', 'polynomial', and 'highest common factor'; and the technique of the Euclidean Algorithm for finding a highest common factor of two polynomials. The reader who has encountered these ideas already may safely omit this chapter.

We shall assume that the reader is familiar with the factorization properties of the integers; and we shall assume a knowledge of ring theory as far as quotient rings. These topics are dealt with in detail in Hartley and Hawkes [14]. As well as summarizing the basic ideas we take the opportunity to introduce some standard notation and terminology.

General properties of rings

We recall that a *ring* is a set R equipped with two binary operations $+$, (addition), and $\times$, (multiplication), such that R is an abelian group under addition, multiplication is associative, and the distributive laws

$$(a+b) \times c = (a \times c) + (b \times c)$$

$$a \times (b+c) = (a \times b) + (a \times c)$$

hold for all a, b, c, $\in R$. The additive identity of R will be denoted by the symbol 0 irrespective of the ring R. We shall write ab instead of $a \times b$.

An *integral domain* is a ring D with three further properties:

(a) Multiplication in D is commutative,

(b) There exists an element $1 \in D$ such that $a1 = 1a = a$ for all $a \in D$,

(c) If $ab = 0$ for $a, b \in D$ then either $a = 0$ or $b = 0$.

A *field* is a ring F such that $F \backslash \{0\}$ is an abelian group under multiplication. Thus every non-zero element $a \in F$ has a multiplicative inverse a^{-1}. We shall often write a/b or $\frac{a}{b}$ instead of ab^{-1}. Every field is an integral domain, the element 1 being the identity element of the multiplicative group.

Important examples of such structures are the integral domain $\mathbf{Z}$ of integers and the fields $\mathbf{Q}$, $\mathbf{R}$, $\mathbf{C}$ of rational, real, and complex numbers (respectively).

A *subring* of a ring R is a non-empty subset S such that if $a, b \in S$ then $a+b \in S$, $a-b \in S$, and $ab \in S$. A *subfield* of a field F is a subset S containing the elements 0 and 1 such that if $a, b \in S$ then $a+b$, $a-b$, $ab \in S$, and further if $a \neq 0$ then $a^{-1} \in S$. An *ideal* of a ring R is a subring I such that if $i \in I$ and $r \in R$ then ir and ri lie in I. Thus $\mathbf{Z}$ is a subring of $\mathbf{Q}$, and $\mathbf{R}$ is a subfield of $\mathbf{C}$, while the set $2\mathbf{Z}$ of even integers is an ideal of $\mathbf{Z}$.

If I is an ideal of the ring R we may form the *quotient ring* R/I which consists of the cosets of I in R (considered as a group under addition) with operations

$$(I+r)+(I+s) = I+(r+s)$$

$$(I+r)(I+s) = I+(rs)$$

where $r, s \in R$ and $I+r$ is the coset $\{i+r : i \in I\}$. For example, let $n\mathbf{Z}$ be the set of integers divisible by a fixed integer n. This is an ideal of $\mathbf{Z}$, and the quotient ring $\mathbf{Z}_n = \mathbf{Z}/n\mathbf{Z}$ is the *ring of integers modulo n* (or *mod n*).

We shall need the following property of $\mathbf{Z}_n$:

1.1 Theorem. *The ring $\mathbf{Z}_n$ is a field if and only if n is a prime number.*

Proof. First suppose n is not prime. If $n = 1$ then $\mathbf{Z}_n = \mathbf{Z}/\mathbf{Z}$ which has only one element and so cannot be a field. If $n > 1$ then $n = rs$ where r and s are integers less than n. Putting $I = n\mathbf{Z}$ we have

$$(I+r)(I+s) = I+rs = I.$$

But I is the zero element of $\mathbf{Z}/I$, while $I+r$ and $I+s$ are non-zero. Since in a field the product of two non-zero elements is non-zero, $\mathbf{Z}/I$ cannot be a field.

Now suppose n is prime. Let $I+r$ be a non-zero element of $\mathbf{Z}/I$. Since r and n are coprime there exist integers a and b such that $ar+bn = 1$ (by standard properties of the integers). Then

$$(I+a)(I+r) = (I+1)-(I+n)(I+b) = I+1$$

and similarly

$$(I+r)(I+a) = I+1.$$

Since $I+1$ is the identity element of $\mathbf{Z}/I$ we have found a multiplicative inverse for the given element $I+r$. Thus every non-zero element of $\mathbf{Z}/I$ has an inverse, so that $\mathbf{Z}_n = \mathbf{Z}/I$ is a field.

From now on when dealing with $\mathbf{Z}_n$ we shall adopt the usual convention and write the elements as $0, 1, 2, \cdots, n-1$ rather than $I, I+1, I+2, \cdots, I+n-1$.

The characteristic of a field

Definition. The *prime subfield* of a field K is the intersection of all subfields of K.

It is easy to see that the intersection of any collection of subfields of K is a subfield (the intersection is not empty since every subfield contains 0 and 1), and therefore the prime subfield of K is the unique smallest subfield of K. Now the fields $\mathbf{Q}$ and $\mathbf{Z}_p$ (p prime) have no proper subfields, so are equal to their prime subfields. The next theorem shows that these are the only fields which can occur as prime subfields.

1.2 Theorem. *Every prime subfield is isomorphic either to the field* $\mathbf{Q}$ *of rationals or the field* $\mathbf{Z}_p$ *of integers modulo a prime number p.*

Proof. Let K be a field, P its prime subfield. P contains 0 and 1, and therefore contains the elements n^* ($n \in \mathbf{Z}$) defined by

$$n^* = 1+1+\cdots+1 \ (n \text{ times}) \text{ if } n > 0$$

$$0^* = 0$$

$$n^* = -(-n)^* \quad \text{if} \quad n < 0.$$

A short calculation using the distributive law shows that the map $*: \mathbf{Z} \to P$ so defined is a ring homomorphism. Two distinct cases arise.

Case 1. $n^* = 0$ for some $n \neq 0$. Since also $(-n)^* = 0$ there exists a smallest positive integer p such that $p^* = 0$. If p were composite, say $p = rs$ where r and s are smaller positive integers, then we should have $r^*s^* = p^* = 0$, so that either $r^* = 0$ or $s^* = 0$, contrary to the definition of p. Therefore p is prime. The elements n^* form a ring isomorphic to $\mathbf{Z}_p$, which is a field by 1.1. This must be the whole of P since P is the smallest subfield of K.

Case 2. $n^* \neq 0$ if $n \neq 0$. Then P must contain all the elements m^*/n^* where m, n are integers and $n \neq 0$. These form a subfield isomorphic to $\mathbf{Q}$ (by the map which sends m^*/n^* to m/n) which is necessarily the whole of P.

Definition. If the prime subfield of K is isomorphic to $\mathbf{Q}$ we say K has *characteristic* 0. If the prime subfield of K is isomorphic to $\mathbf{Z}_p$ we say K has *characteristic p*.

For example, the fields **Q**, **R**, **C**, all have characteristic zero since in each case the prime subfield is **Q**. The field $\mathbf{Z}_p$ (p prime) has characteristic p. We shall see later that there are other fields of characteristic p.

The elements n^* defined in the proof of the above theorem are of considerable importance in what follows. It is conventional to omit the asterisk and write simply n instead of n^*. This abuse of notation will cause no confusion as long as it is understood that n may be zero in the field without being zero as an integer. Thus in $\mathbf{Z}_5$ we have $2 = 7 = -3$, *etc.* This difficulty does not arise in fields of characteristic zero.

With this convention a product nk ($n \in \mathbf{Z}$, $k \in K$) makes sense, and we have

$$nk = \pm(k + \cdots + k).$$

1.3 Lemma. *If K is a subfield of L then K and L have the same characteristic.*

Proof. In fact, K and L have the same prime subfield.

1.4 Lemma. *If k is a non-zero element of the field K, and if n is an integer such that $nk = 0$, then n is a multiple of the characteristic of K.*

Proof. We must have $n = 0$ in K, i.e. in old notation $n^* = 0$. If the characteristic is 0 this implies $n = 0$ (as an integer). If the characteristic is $p > 0$ this implies that n is a multiple of p.

Fields of fractions

Sometimes it is possible to embed a ring R in a field; that is, find a field containing a subring isomorphic to R. Thus **Z** can be embedded in **Q**. This particular example has the property that every element of **Q** is a fraction whose numerator and denominator lie in **Z**. We wish to generalize this situation.

Definition A *field of fractions* of the ring R is a field K containing a subring R' isomorphic to R, such that every element of K can be expressed in the form r/s for $r, s \in R'$, where $s \neq 0$.

An analysis of the way in which $\mathbf{Z}$ is embedded in $\mathbf{Q}$ leads us to state:

1.5 Theorem. *Every integral domain possesses a field of fractions.*

Proof. Let R be an integral domain, and let S be the set of all ordered pairs (r, s) where r and s lie in R and $s \neq 0$. Define a relation $\sim$ on S by

$$(r, s) \sim (t, u) \Leftrightarrow ru = st.$$

It is easy to verify that $\sim$ is an equivalence relation; we denote the equivalence class of (r, s) by $[r, s]$. The set F of equivalence classes will provide the required field of fractions. First we define the operations on F by

$$[r, s] + [t, u] = [ru + ts, su]$$
$$[r, s][t, u] = [rt, su].$$

Then we perform a long series of computations to show that F has all the required properties. Since these computations are routine we shall not perform them here, but any reader who has never done so should check them for himself. The following must be demonstrated:

(a) The operations are well-defined. That is to say, if $(r, s) \sim (r', s')$ and $(t, u) \sim (t', u')$, then $[r, s] + [t, u] = [r', s'] + [t', u']$ and $[r, s][t, u] = [r', s'][t', u']$.

(b) They are operations on F (this is where we need to know that R is an integral domain).

(c) F is a field.

(d) The map $R \to F$ which sends $r \to [r, 1]$ is a monomorphism.

(e) $[r, s] = [r, 1]/[s, 1]$.

It can be shown (see Exercise 1.16) that for a given integral domain R, all fields of fractions are isomorphic. We can therefore refer to the field constructed above as *the* field of fractions of R. It is customary to identify an element $r \in R$ with its image $[r, 1]$ in F; whereupon we have $[r, s] = r/s$.

Polynomials.

There is considerable confusion as to what a polynomial is. A precise, logical definition can easily be given (see Hartley and Hawkes [14] p. 37) but we prefer not to be too formal. Let R be a ring in which multiplication is commutative. We shall define a *polynomial over R in the indeterminate t* to be an expression

$$r_0 + r_1 t + \cdots + r_n t^n$$

where $r_0, \cdots, r_n \in R$, $0 \leq n \in \mathbf{Z}$, and t is undefined. The elements $r_0, \cdots, r_n$ are said to be the *coefficients* of the polynomial. In the usual way terms $0t^m$ may be omitted or written as 0, and $1t^m$ can be replaced by t^m.

Two polynomials are *defined* to be equal if and only if the corresponding coefficients are equal (with the understanding that powers of t not occurring in the polynomial may be taken to have zero coefficient). Failure to observe this convention is the cause of most of the confusion alluded to above. The sum and the product of two polynomials are *defined* as follows: write

$$\sum r_i t^i$$

instead of

$$r_0 + r_1 t + \cdots + r_n t^n$$

where the summation is considered as being over all integers $i \geq 0$, and r_k is defined to be 0 if $k \geq n$. Then, if

$$r = \sum r_i t^i$$
$$s = \sum s_i t^i$$

we define

$$r+s = \sum (r_i + s_i)t^i$$
$$rs = \sum q_j t^j$$

where

$$q_j = \sum_{h+i=j} r_h s_i.$$

It is now easy to check directly from these definitions that the set of all polynomials over R in the indeterminate t is a ring – the *ring of polynomials over R in the indeterminate t*. We denote this by the symbol $R[t]$. For different indeterminates u, v, w, $\cdots$ we obtain polynomial rings $R[u]$, $R[v]$, $R[w]$, $\cdots$ which are all isomorphic in an obvious manner. We can also define polynomials in *several* indeterminates t_1, t_2, $\cdots$ and obtain the polynomial ring

$$R[t_1, t_2, \cdots],$$

in an analogous way.

An element of $R[t]$ will usually be denoted by a single letter, such as f, whenever it is clear which indeterminate is involved. If there is ambiguity we shall write $f(t)$ to emphasize the role played by t. This notation has the unfortunate effect of making f *appear* to be a function, with t as a 'variable'. This is not so. However, each polynomial $f \in R[t]$ defines a function from R to R, which acts in the following way: if

$$f = \sum r_i t^i$$

and $\alpha \in R$ then α is mapped to

$$\sum r_i \alpha^i.$$

This latter is simply an element of R, *not* a polynomial. It is customary to use the same symbol, f, to denote this function; so we have $f(\alpha) = \Sigma\, r_i \alpha^i$. Again the abuse of language is traditional, and with care should cause no confusion.

A related abuse of notation allows us to 'change the

variable' in a polynomial. Thus if t, u are two indeterminates and $f(t) = \Sigma r_i t^i$ we may define $f(u) = \Sigma r_i u^i$. It is also clear what is meant by, say, $f(t+1)$, etc.

The main point to bear in mind here is that two distinct polynomials over R may give rise to the same function on R. For example, over $\mathbf{Z}_2$ the polynomials t and t^2 both give rise to the identity map. This is one reason why it is unwise to define polynomials as functions. Another reason is that we wish to talk of one polynomial *dividing* another; but this concept makes little sense for functions on R, especially when R happens to be a field.

When R is an integral domain $R[t]$ has a field of fractions. This follows from:

1.6 Lemma. *If R is an integral domain and t is an indeterminate then $R[t]$ is an integral domain.*

Proof. Suppose that $f = f_0 + f_1 t + \cdots + f_n t^n$, $g = g_0 + g_1 t + \cdots + g_m t^m$ where $f_n \neq 0 \neq g_m$ and all the coefficients lie in R. The coefficient of t^{m+n} in fg is $f_n g_m$, which is non-zero since R is an integral domain. Thus if f, g are non-zero then fg is non-zero. This implies that $R[t]$ is an integral domain as claimed.

Theorem 1.5 now implies that $R[t]$ has a field of fractions, which we call the *field of rational expressions in t over R* and denote by $R(t)$. Similarly $R[t_1, \cdots, t_n]$ has a field of fractions $R(t_1, \cdots, t_n)$.

The Euclidean Algorithm

We shall need the:

Definition. If f is a polynomial over a commutative ring R and $f \neq 0$ then the *degree* of f is the highest power of t occurring in f with non-zero coefficient.

That is to say, if $f = \Sigma r_i t^i$ and $r_n \neq 0$ and $r_m = 0$ for $m > n$, then f has degree n. We write ∂f for the degree of f.

To deal with the case $f = 0$ we adopt the not very happy convention that $\partial 0 = -\infty$ (which symbol is endowed with the following properties: $-\infty < n$ for any integer n, $-\infty + n = -\infty$, $-\infty . n = -\infty$, $(-\infty)^2 = -\infty$).

The following results are immediate from this definition:

1.7 Proposition. *If R is an integral domain and f, g are polynomials over R, then*

$$\partial(f+g) \leq \max(\partial f, \partial g)$$
$$\partial(fg) = \partial f + \partial g.$$

(The inequality in the first line is due to the possibility of the highest terms 'cancelling'.)

Many important results in the factorization theory of polynomials derive from the observation that one polynomial may always be divided by another provided that a 'remainder' term is allowed.

1.8 Proposition. *Let f and g be polynomials over a field K, and suppose that f is non-zero. Then there exist unique polynomials q and r over K such that*

$$g = fq + r$$

where r has smaller degree than f.

Proof. We use induction on the degree of g. If $\partial g = -\infty$ then $g = 0$ and we may take $q = r = 0$. If $\partial g = 0$ then $g = k \in K$. If also $\partial f = 0$ then $f = j \in K$ and we may take $q = k/j$ and $r = 0$. Otherwise $\partial f > 0$ and we may take $q = 0$ and $r = g$. This starts the induction.

Now assume the result holds for all polynomials of degree $< n$, and let $\partial g = n > 0$. If $\partial f > \partial g$ we may as before take $q = 0$, $r = g$. Otherwise we have

$$f = a_m t^m + \cdots + a_0$$
$$g = b_n t^n + \cdots + b_0$$

where $a_m \neq 0 \neq b_n$ and $m \leq n$. Now put

$$g_1 = b_n a_m^{-1} t^{n-m} f - g.$$

Since terms of highest degree cancel (which is the object of the exercise!) we have $\partial g_1 < \partial g$. By induction we can find polynomials q_1 and r_1 such that

$$g_1 = f q_1 + r_1$$

and $\partial r_1 < \partial f$. If we now let

$$q = b_n a_m^{-1} t^{n-m} - q_1$$

$$r = r_1$$

we find that

$$g = fq + r$$

and $\partial r < \partial f$, as required.

Finally we prove uniqueness. Suppose that

$$g = f q_1 + r_1 = f q_2 + r_2 \quad \text{where} \quad \partial r_1, \partial r_2 < \partial f.$$

Then

$$f(q_1 - q_2) = r_1 - r_2.$$

By 1.7 the polynomial on the left has higher degree than that on the right, unless both are zero. Since $f \neq 0$ we must have $q_1 = q_2$ and $r_1 = r_2$. Thus q and r are unique.

With the above notation q is called the *quotient* and r is called the *remainder* (on dividing g by f). The process of finding q and r is sometimes called the *Division Algorithm*.

Our next step is to introduce notions of divisibility for polynomials, and in particular the idea of 'highest common factor' which is crucial to the 'arithmetic' of polynomials in the next chapter.

Definition. Let f and g be polynomials over a field K. We say that f *divides* g (or f *is a factor of* g, or g *is a multiple of* f) if there exists some polynomial h over K such that $g = fh$.

The notation

$$f \mid g$$

will mean that f divides g, while

$$f \nmid g$$

will mean that f does not divide g. (These notations are due to Gauss.)

A polynomial d over K is a *highest common factor* (*hcf*) of f and g if $d \mid f$ and $d \mid g$ and further, whenever $e \mid f$ and $e \mid g$ we have $e \mid d$.

Note that we have said *a* highest common factor rather than *the* highest common factor. This is because hcf's need not be unique. The next lemma shows that they are unique *apart from constant factors* (where 'constant' means a polynomial of degree 0 – this terminology is historical).

1.9 Lemma. *If d is an hcf of the polynomials f and g over a field K, and if $0 \neq k \in K$, then kd is also an hcf for f and g.*

If d and e are two hcf's for f and g then there exists a non-zero element $k \in K$ such that $e = kd$.

Proof. Clearly $kd \mid f$ and $kd \mid g$. If $e \mid f$ and $e \mid g$ then $e \mid d$ so that $e \mid kd$. Hence kd is an hcf.

If d and e are hcf's then by definition $e \mid d$ and $d \mid e$. Thus $e = hd$ for some polynomial h. Since $e \mid d$ the degree of e is less than or equal to the degree of d, so h must have degree ≤ 0. Therefore $h = k \in K$. Since $0 \neq e = kd$, we must have $k \neq 0$.

We shall prove that any two non-zero polynomials over a field have at least one hcf by providing a method by which an hcf may be calculated. This is a generalization of one given by Euclid (c. 600 B.C.) for calculating hcf's of integers, and is accordingly known as the *Euclidean Algorithm*.

1.10 Algorithm.
Ingredients. Two polynomials f and g, both non-zero, over a field K.

Recipe. For notational convenience let $f = r_{-1}$, $g = r_0$. Using the division algorithm find successively polynomials q_i and r_i over K such that

$$r_{-1} = q_1 r_0 + r_1 \qquad \partial r_1 < \partial r_0$$
$$r_0 = q_2 r_1 + r_2 \qquad \partial r_2 < \partial r_1$$
$$r_1 = q_3 r_2 + r_3 \qquad \partial r_3 < \partial r_2 \qquad (*)$$
$$\cdots$$
$$r_i = q_{i+2} r_{i+1} + r_{i+2} \quad \partial r_{i+2} < \partial r_{i+1}$$
$$\cdots$$

Since the degrees of the r_i decrease, we must eventually reach a point where the process stops; and this can only happen if some $r_{s+2} = 0$. The last equation in the list then reads

$$r_s = q_{s+2} r_{s+1}. \qquad (**)$$

1.11 Theorem. *With the above notation r_{s+1} is an hcf for f and g over K.*

Proof. First we show that r_{s+1} divides both f and g. We use descending induction to show that $r_{s+1} \mid r_i$ for all i. Clearly $r_{s+1} \mid r_{s+1}$. By (**) $r_{s+1} \mid r_s$. Using (*) we see that if $r_{s+1} \mid r_{i+2}$ and $r_{s+1} \mid r_{i+1}$ then $r_{s+1} \mid r_i$. Hence $r_{s+1} \mid r_i$ for all i; in particular $r_{s+1} \mid r_0 = g$ and $r_{s+1} \mid r_{-1} = f$.

Now suppose that $e \mid f$ and $e \mid g$. From (*) it follows inductively that $e \mid r_i$. Hence $e \mid r_{s+1}$. Therefore r_{s+1} is an hcf for f and g as claimed.

Example. Let $f = 2t^7 + t^3 - 1$, $g = t^3 + 4$ over **Q**. We compute an hcf as follows:

$$2t^7 + t^3 - 1 = (t^3 + 4)(2t^4 - 8t + 1) + (-32t - 5).$$

$$t^3 + 4 = (32t - 5)\left(\frac{t^2}{32} + \frac{5t}{1024} - \frac{25}{32768}\right) + \frac{130937}{32768}$$

$$32t - 5 = \frac{130937}{32768}\left(\frac{32768}{130937}(32t - 5)\right) + 0.$$

Hence $\dfrac{130937}{32768}$ is an hcf. So is any rational multiple of it –
in particular 1!

We end this chapter by deducing from the Euclidean Algorithm an important property of the hcf of two polynomials.

1.12 Theorem. *Let f and g be non-zero polynomials over a field K and let d be an hcf for f and g. Then there exist polynomials a and b over K such that*

$$d = af + bg.$$

Proof. Since hcf's are unique up to constant factors we may assume that $d = r_{s+1}$ where (*) and (**) hold. We claim as induction hypothesis that there exist polynomials a_i and b_i such that

$$d = a_i r_i + b_i r_{i+1}.$$

This is clearly true when $i = s+1$, for we may then take $a_i = 1$, $b_i = 0$. Now from (*)

$$r_{i+1} = r_{i-1} - q_{i+1} r_i.$$

Hence by induction

$$d = a_i r_i + b_i (r_{i-1} - q_{i+1} r_i)$$

so that if we put

$$a_{i-1} = b_i$$

$$b_{i-1} = a_i - b_i q_{i+1}$$

we have

$$d = a_{i-1} r_{i-1} + b_{i-1} r_i.$$

Hence by descending induction

$$d = a_{-1} r_{-1} + b_{-1} r_0$$

$$= af + bg$$

where $a = a_{-1}$, $b = b_{-1}$. This completes the proof.

The induction step above affords a practical method of calculating a and b in any particular case.

Exercises

1.1 Show that $15\mathbf{Z}$ is an ideal of $5\mathbf{Z}$, and that $5\mathbf{Z}/15\mathbf{Z}$ is isomorphic to $\mathbf{Z}_3$.

1.2 Are the rings $\mathbf{Z}$ and $2\mathbf{Z}$ isomorphic?

1.3 Write out addition and multiplication tables for $\mathbf{Z}_6$, $\mathbf{Z}_7$, and $\mathbf{Z}_8$. Which of these rings are integral domains? Which are fields?

1.4 Why are the addition and multiplication tables of exercise 1.3 symmetric about the diagonal?

1.5 Define a *prime field* to be a field with no proper subfields. Show that the prime fields (up to isomorphism) are precisely $\mathbf{Q}$ and $\mathbf{Z}_p$ (p prime).

1.6 Show that the following tables define a field.

+	0	1	α	β		.	0	1	α	β
0	0	1	α	β		0	0	0	0	0
1	1	0	β	α		1	0	1	α	β
α	α	β	0	1		α	0	α	β	1
β	β	α	1	0		β	0	β	1	α

Find its prime subfield and its characteristic. Is it isomorphic to $\mathbf{Z}_4$? How many fields are there (up to isomorphism) with exactly 4 elements?

1.7 Check all the details omitted from the proof of theorem 1.5.

1.8 By considering the polynomials $3t$ and $2t$ over $\mathbf{Z}_6$ show that the second part of proposition 1.7 fails if R is a ring rather than an integral domain. What happens to the first part of the proposition under these circumstances?

1.9 For the following pairs of polynomials f and g find the quotient and remainder on dividing g by f.
(a) $g = t^7 - t^3 + 5, f = t^3 + 7$ over $\mathbf{Q}$.
(b) $g = t^2 + 1, f = t^2$ over $\mathbf{Q}$.
(c) $g = 4t^3 - 17t^2 + t - 3, f = 2t + 5$ over $\mathbf{R}$.
(d) $g = t^3 + 2t^2 - t + 1, f = t + 2$ over $\mathbf{Z}_3$.
(e) $g = t^7 - 4t^6 + t^3 - 3t + 5, f = 2t^3 - 2$ over $\mathbf{Z}_7$.

1.10 Find hcf's for these pairs of polynomials, and check that your results are common factors of f and g.

1.11 Express these hcf's in the form $af + bg$.

1.12 An element x of a ring R with multiplicative identity 1 is said to be a *unit* of R if it has a multiplicative inverse. Show that the units of R form a group under multiplication.

1.13 Find the structure of the group of units of $\mathbf{Z}_5$, $\mathbf{Z}_6$, $\mathbf{Z}_{12}$, $\mathbf{Z}_{24}$.

1.14 For what values of n does every element of the group of units of $\mathbf{Z}_n$ have order dividing 2?

1.15 Mark the following true or false.
(a) I have come across everything in this chapter before.
(b) Any pair of polynomials has an hcf.
(c) The function ∂ which sends f to ∂f is a ring homomorphism $R[t] \to \mathbf{Z}$.
(d) Every ring has a field of fractions.
(e) Every field is isomorphic to its field of fractions.

(f) $\mathbf{Z}_n$ is an integral domain if and only if it is a field.

(g) Every integral domain is a field.

(h) Every field is an integral domain.

(i) A polynomial over K is a function $K \to K$.

(j) If K is a field then $K[t]$ is a field.

1.16 Let D be an integral domain with a field of fractions F. Let K be any field. Prove that any monomorphism $\phi : D \to K$ has a unique extension to a monomorphism $\psi : F \to K$ defined by

$$\psi(a/b) = \phi(a)/\phi(b)$$

for $a, b \in D$. By considering the case where K is another field of fractions for D and ϕ is the inclusion map show that fields of fractions are unique up to isomorphism.

CHAPTER TWO

Factorization
of polynomials

Polynomial equations $f(\alpha) = 0$ (where f is a polynomial) have been of considerable historical importance in mathematics. For any given polynomial f the problem of finding a solution, or even all of them, is one of computation; but a general theory requires more refined techniques. It was noticed early on that if f is a product gh of polynomials of smaller degree, then the solutions of $f(\alpha) = 0$ are precisely those of $g(\alpha) = 0$ together with those of $h(\alpha) = 0$. From this simple idea emerged the *arithmetic* of polynomials; a systematic study of divisibility properties of polynomials with particular reference to analogies with the integers. The Euclidean Algorithm of the preceding chapter is a case in point.

In this chapter we define the relevant notions of divisibility and show that there are certain polynomials, the 'irreducible' ones, which play a similar role to prime numbers in the ring of integers. Every polynomial over a field can be expressed as a product of irreducible polynomials in an essentially unique way. We also define 'zeros' of polynomials and relate these to the factorization theory. In the final section we investigate how to reconstruct a polynomial from its zeros.

Irreducibility

The following definition is not quite the same as that given in Hartley and Hawkes [14], but it is more suitable for our purposes. For polynomials over a *field* it coincides with their definition.

Definition. A polynomial over a commutative ring is *reducible* if it is a product of two polynomials of smaller degree. Otherwise it is *irreducible*.

Examples.

1 All polynomials of degree 0 or 1 are irreducible, since they certainly cannot be expressed as a product of polynomials of *smaller* degree.

2 The polynomial $t^2 - 2$ is irreducible over $\mathbb{Q}$. To show this we suppose, for a contradiction, that it is reducible. Then

$$t^2 - 2 = (at + b)(ct + d)$$

where a, b, c, $d \in \mathbb{Q}$. Dividing out if necessary we may assume $a = c = 1$. Then $b + d = 0$ and $bd = -2$, so that $b^2 = 2$. But no *rational* number has its square equal to 2.

3 However, $t^2 - 2$ is reducible over $\mathbb{R}$, for now

$$t^2 - 2 = (t - \sqrt{2})(t + \sqrt{2}).$$

This shows that an irreducible polynomial may become reducible over a larger field.

Any reducible polynomial can be written as the product of two polynomials of smaller degree. If either of these is reducible it too can be split up into factors of smaller degree . . . and so on. This process must terminate since the degrees cannot decrease indefinitely. This is the idea behind the proof of:

2.1 Theorem. *Any non-zero polynomial over a field K is a product of irreducible polynomials over K.*

Proof. Let g be any non-zero polynomial over K. We proceed by induction on the degree of g. If $\partial g = 0$ or 1 then g is automatically irreducible. If $\partial g > 1$ either g is irreducible or $g = hj$ where ∂h, $\partial j < \partial g$. By induction h and j are products of irreducible polynomials, whence g is such a product. The theorem follows by induction.

The importance of prime numbers in **Z** stems not so much from the possibility of factorizing every integer into primes, but more from the uniqueness (up to order) of the prime factors. Likewise the importance of irreducible polynomials depends upon a uniqueness theorem. Uniqueness of factorization is *not* obvious (see Hardy and Wright [13] p. 211). In certain rings it is possible to express every element as a product of irreducible elements, without this expression being in any way unique. Kummer made the error of assuming uniqueness of factorization in an attempt to prove Fermat's Last Theorem‡, and his efforts to put things right resulted in the creation of a new branch of mathematics – the theory of ideals in algebraic number fields. We shall heed the warning and *prove* the uniqueness of factorization for polynomials over a field.

For convenience we make the following:

Definition. If f and g are polynomials over a field K with hcf equal to 1, we say that f and g are *coprime* (or f *is prime to* g).

Next we prove a lemma.

2.2 Lemma.   *If K is a field, f is an irreducible polynomial over K, and g, h are polynomials over K such that f divides gh, then either f divides g or f divides h.*

Proof. Suppose that $f \nmid g$. We claim that f and g are coprime. For if d is an hcf for f and g, then since f is irreducible and $d \mid f$ either $d = kf (k \in K)$ or $d = k \in K$. In the first case

‡   The equation $x^n + y^n = z^n$ has no non-zero integer solutions if n is an integer ≥ 3. This is known to be true if $n \leq 5500$ (see Kobelev [17]) but a general proof is lacking. For a historical view of the problem see Mordell [45].

we have $f \mid g$ contrary to hypothesis. In the second case 1 is also an hcf for f and g, so that they are coprime. By 1.12 there exist polynomials a and b over K such that

$$1 = af + bg.$$

Then

$$h = haf + hbg.$$

Now $f \mid haf$, and $f \mid hbg$ since $f \mid gh$. Hence $f \mid h$. This completes the proof.

We may now prove the uniqueness theorem.

2.3 Theorem. *For any field K, factorization of polynomials over K into irreducible polynomials is unique up to constant factors and the order in which the factors are written.*

Proof. Suppose that $f = f_1 \cdots f_r = g_1 \cdots g_s$ where f is a polynomial over K and $f_1, \cdots, f_r, g_1, \cdots, g_s$ are irreducible polynomials over K. If all the f_i are constant then $f \in K$ so that all the g_j are constant. Otherwise we may assume that *no f_i is constant* (by dividing out all the constant terms). Then $f_1 \mid g_1 \cdots g_s$. By an obvious induction based on 2.2, $f_1 \mid g_i$ for some i. We can choose notation so that $i = 1$, and then $f_1 \mid g_1$. Since f_1 and g_1 are irreducible and f_1 is not a constant we must have $f_1 = k_1 g_1$ for some constant k_1. Similarly $f_2 = k_2 g_2, \cdots, f_r = k_r g_r$ where $k_2, \cdots, k_r$ are constant. The remaining g_j ($j > r$) must also be constant, or else the degree of the right-hand side would be too large. The result follows.

Tests for irreducibility

It is in general very difficult to tell whether or not a given polynomial is irreducible. As an example, think about

$$t^{16} + t^{15} + t^{14} + t^{13} + t^{12} + t^{11} + t^{10} + t^9 + t^8 + t^7 + t^6 + t^5 + t^4 + t^3 + t^2 + t + 1.$$

(This is not an idle example: we shall be considering precisely this polynomial in chapter 17.)

To try all possible factors is usually futile, if only because there are *a priori* infinitely many possibilities; though with enough short-cuts this is a method that can be used if all else fails! What we want is a repertory of simple tricks which can be resorted to in time of need. The most important of these is *Eisenstein's Criterion*. It applies in the first instance to polynomials over **Z**. However, it is known that irreducibility over **Z** is equivalent to irreducibility over **Q** (proved by Gauss). We prove this result first.

2.4 Proposition. *Let f be a polynomial over* **Z** *which is irreducible over* **Z**. *Then f, considered as a polynomial over* **Q**, *is also irreducible over* **Q**.

Proof. The point of this proposition is that when we extend to **Q** there are hosts of new polynomials which, it appears, might be factors of f. We shall show that in fact they are not. So we suppose that f is irreducible over **Z** but reducible over **Q**, so that $f = gh$ where g and h are polynomials over **Q**, of smaller degree. Multiplying through by the product of the denominators of the coefficients of g and h we can rewrite this in the form

$$nf = g'h'$$

where $n \in$ **Z** and g', h' are polynomials over **Z**. We now show that we can cancel out the prime factors of n one by one, without going outside **Z**$[t]$.

Suppose that p is a prime factor of n. We claim that, if

$$g' = g_0 + g_1 t + \cdots + g_r t^r$$
$$h' = h_0 + h_1 t + \cdots + h_s t^s,$$

then either p divides all the coefficients g_i or else p divides all the coefficients h_j. If not, there must be *smallest* values i and j such that $p \nmid g_i$ and $p \nmid h_j$. However p divides the coefficient of t^{i+j} in $g'h'$, which is

$$h_0 g_{i+j} + h_1 g_{i+j-1} + \cdots + h_j g_i + \cdots + h_{i+j} g_0$$

and by the choice of i and j, the prime p divides every term of this expression except perhaps $h_j g_i$. But p divides the whole expression, so $p \,|\, h_j g_i$. But $p \nmid h_j$ and $p \nmid g_i$, a contradiction. This establishes our claim.

Without loss of generality we may assume that p divides every coefficient g_i. Then $g' = pg''$ where g'' is a polynomial over $\mathbf{Z}$ of the same degree as g' (or g). Let $n = pn_1$. Then

$$pn_1 f = pg''h'$$

so that

$$n_1 f = g''h'.$$

Proceeding in this way we can remove all the prime factors of n, arriving at an equation

$$f = \bar{g}\bar{h}$$

where $\bar{g}$ and $\bar{h}$ are polynomials over $\mathbf{Z}$ which are rational multiples of the original g, h. But this contradicts the irreducibility of f over $\mathbf{Z}$.

Hence our assumption of reducibility over $\mathbf{Q}$ is false, and f must be irreducible over $\mathbf{Q}$.

Thus armed, we can prove:

2.5 Theorem. (*Eisenstein's Irreducibility Criterion*). *Let*

$$f(t) = a_0 + a_1 t + \cdots + a_n t^n$$

be a polynomial over $\mathbf{Z}$. *Suppose that there is a prime q such that*

(1) $q \nmid a_n$
(2) $q \,|\, a_i \, (i = 0, \cdots, n-1)$
(3) $q^2 \nmid a_0$.

Then f is irreducible over $\mathbf{Q}$.

Proof. By 2.4 it suffices to show that f is irreducible over $\mathbf{Z}$. Suppose for a contradiction that $f = gh$ where

$$g = b_0 + b_1 t + \cdots + b_r t^r$$

$$h = c_0 + c_1 t + \cdots + c_s t^s$$

are polynomials of smaller degree over $\mathbf{Z}$. Then $r+s = n$. Now $b_0 c_0 = a_0$ so by (2) $q \,|\, b_0$ or $q \,|\, c_0$. By (3) q cannot divide both b_0 and c_0, so without loss of generality we can assume $q \,|\, b_0$, $q \nmid c_0$. If all coefficients b_i were divisible by q then a_n would be divisible by q, contrary to (1). Let b_i be the first coefficient of g not divisible by q. Then

$$a_i = b_i c_0 + \cdots + b_0 c_i$$

where $i < n$. This implies that q divides c_0, since q divides $a_i, b_0, \cdots, b_{i-1}$, but not b_i. This is a contradiction. Hence f is irreducible.

Examples.
1 Consider $f(t) = \frac{2}{9}t^5 + \frac{5}{3}t^4 + t^3 + \frac{1}{3}$ over $\mathbf{Q}$. This is irreducible if and only if $9f(t) = 2t^5 + 15t^4 + 9t^3 + 3$ is irreducible over $\mathbf{Q}$. Eisenstein's criterion applies with $q = 3$, showing that f is irreducible.
2 Consider $f(t) = t^{16} + t^{15} + \cdots + 1$ as mentioned above. As it stands f is not susceptible to treatment. But $f(t)$ is obviously irreducible if and only if $f(t+1)$ is irreducible. If we expand this monster it transpires that Eisenstein's criterion applies with $q = 17$, so that f is irreducible over $\mathbf{Q}$. (There is, incidentally, a good reason why this method works which does not emerge clearly from direct calculations. See lemma 17.9 for details.)

There is another very useful test for irreducibility which is most easily explained by an example. The idea is this: the natural homomorphism $\mathbf{Z} \to \mathbf{Z}_n$ extends in an obvious way to a homomorphism $\mathbf{Z}[t] \to \mathbf{Z}_n[t]$. Now a reducible polynomial over $\mathbf{Z}$ is a product gh of polynomials of lower degree; and this factorization is preserved by the homomorphism. Provided n does not divide the highest coefficient of the given polynomial the image is reducible over $\mathbf{Z}_n$. So if the image of a polynomial is irreducible over $\mathbf{Z}_n$ then the original polynomial must be irreducible over $\mathbf{Z}$. *Since $\mathbf{Z}_n$ is finite there are only finitely many possibilities to check when deciding irreducibility.*

In practice the trick is to choose the right value for n.

Example. Consider $f(t) = t^4 + 15t^3 + 7$ over $\mathbf{Z}$. Over $\mathbf{Z}_5$ this becomes $t^4 + 2$. If this is reducible over $\mathbf{Z}_5$ then either it has a factor of degree 1 or it is a product of two factors of degree 2. The first possibility gives rise to an element $x \in \mathbf{Z}_5$ such that $x^4 + 2 = 0$. No such element exists (there are only 5 elements to check!) so this case is ruled out. In the remaining case we have, without loss of generality,

$$t^4 + 2 = (t^2 + at + b)(t^2 + ct + d).$$

Therefore $a + c = 0, ac + b + d = 0, bd = 2$. Thus $b + d = a^2$, which can take only the values 0, 1, 4 since these are the only squares in $\mathbf{Z}_5$. Hence either $b(1 - b) = 2$ or $-b^2 = 2$ or $b(4 - b) = 2$. Trying all possible values 0, 1, 2, 3, 4 for b we see that none of these equations can hold. Hence $t^4 + 2$ is irreducible over $\mathbf{Z}_5$, and therefore the original $f(t)$ is irreducible over $\mathbf{Z}$, hence over $\mathbf{Q}$.

Notice that if we work in $\mathbf{Z}_3$ then $f(t)$ becomes $t^4 + 1$ which equals $(t^2 + t - 1)(t^2 - t - 1)$ and so is reducible. Thus working mod 3 fails to give us a proof of irreducibility.

These trial and error methods are not entirely satisfactory, and it would be nice to have a method which always works, at any rate in theory, even if the necessary computations are too long to be of practical use. For the field $\mathbf{Q}$ such an algorithm is known, but in general the question is open. The method for $\mathbf{Q}$, due to Kronecker (see Van der Waerden [5] p. 77) could in principle be carried out on a computer. It is not always the case that such *decision procedures* are available for solving problems; the famous theorem of Gödel shows that there is no general recipe by which we can decide whether or not a given statement in ordinary arithmetic is true (where 'statement' and 'recipe' have very precise logical definitions). For details see Mendelson [21], Gödel [10].

Zeros of polynomials

We begin with a formal definition.

Definition. Let R be a commutative ring, and f a polynomial over R. Any element $\alpha \in R$ such that $f(\alpha) = 0$ is a *zero* of f in R.

For the moment we consider polynomials over the real numbers. Here we can draw the graph $y = f(x)$ (in standard terminology) which, in a typical case, resembles Fig. 1.

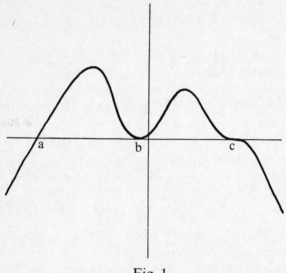

Fig. 1

The zeros of $f(t)$ are the points at which the curve crosses the x-axis. Consider the three zeros marked a, b, c in the diagram. At a the curve cuts straight through the axis; at b it 'bounces' off it; at c it 'slides' through horizontally. These phenomena are generally distinguished by saying that b and c are 'multiple zeros' of $f(t)$. The single zero b must be thought of as two equal zeros (or more) and c as three (or more).

But if they are equal, how can there be two of them?

To clarify our ideas on this question, we must look at *linear* factors (i.e. factors of degree 1) of f.

2.6 Lemma. *Let f be a polynomial over the field K. An* *element $\alpha \in K$ is a zero of f if and only if $(t-\alpha) \mid f(t)$.*

Proof. If $(t-\alpha) \mid f(t)$ then $f(t) = (t-\alpha)g(t)$ for some polynomial g over K, so that $f(\alpha) = (\alpha-\alpha)g(\alpha) = 0$.

Conversely, suppose $f(\alpha) = 0$. By the division algorithm 1.8 there exist polynomials q, r over K such that

$$f(t) = (t-\alpha)q(t) + r(t)$$

where $\partial r < 1$. Thus $r(t) = r \in K$. Substituting α for t we find that

$$0 = f(\alpha) = (\alpha-\alpha)q(\alpha) + r$$

so that $r = 0$. Hence $(t-\alpha) \mid f(t)$.

We can now say what we mean by a multiple zero.

Definition. Let f be a polynomial over the field K. An element $\alpha \in K$ is a *simple* zero of f if $(t-\alpha) \mid f(t)$ but $(t-\alpha)^2 \nmid f(t)$. The element α is a zero of f of *multiplicity m* if $(t-\alpha)^m \mid f(t)$ but $(t-\alpha)^{m+1} \nmid f(t)$. Zeros of multiplicity greater than 1 are *repeated* or *multiple* zeros.

For example, $t^3 - 3t + 2$ over $\mathbf{Q}$ has zeros at $\alpha = 1, -2$. It factorizes as $(t-1)^2(t+2)$. Hence -2 is a simple zero, while 1 is a zero of multiplicity 2.

When $K = \mathbf{R}$ and we draw a graph, as in Fig. 1, points like a are the simple zeros; points like b are zeros of even multiplicity; and points like c are zeros of odd multiplicity > 1. For fields other than $\mathbf{R}$ (except perhaps $\mathbf{Q}$, or other subfields of $\mathbf{R}$) a graph has no meaning; but the simple geometrical picture for the real field is often helpful in understanding what is happening in general. Of course, a picture alone does not comprise a proof.

2.7 Lemma. *Let f be a non-zero polynomial over the field K, and let its distinct zeros be $\alpha_1, \cdots, \alpha_r$ with multiplicities $m_1, \cdots, m_r$ respectively. Then*

$$f(t) = (t-\alpha_1)^{m_1} \cdots (t-\alpha_r)^{m_r} g(t) \tag{1}$$

where g has no zeros in K.

Conversely, if (1) *holds and g has no zeros in K, then the zeros of f in K are* $\alpha_1, \cdots, \alpha_r$ *with multiplicities* $m_1, \cdots, m_r$ *respectively.*

Proof. For any $\alpha \in K$ the polynomial $t - \alpha$ is irreducible. Hence for distinct $\alpha, \beta \in K$ the polynomials $t - \alpha$ and $t - \beta$ are coprime. By uniqueness of factorization (Theorem 2.3) equation (1) must hold; and g cannot have any zeros in K or else f would have extra zeros, or zeros of larger multiplicity.

The converse follows simply from uniqueness of factorization.

From this lemma we can deduce a famous theorem:

2.8 Theorem. *The number of zeros of a polynomial over a field, counted according to multiplicity, is less than or equal to its degree.*

Proof. In equation (1) we must have $m_1 + \cdots + m_r < \partial f$.

Symmetric polynomials

Usually we are given a polynomial and wish to find its zeros. But it is also possible to work in the opposite direction: given the zeros (with multiplicities) reconstruct the polynomial. This is a far easier problem which has a complete general solution, but despite its simplicity it is of considerable theoretical importance.

Consider a polynomial of degree n having its full quota of n zeros (counting multiplicities). It is therefore a product of n linear factors

$$f(t) = k(t - \alpha_1) \cdots (t - \alpha_n)$$

where $k \in K$ and the α_i are the zeros in K (not necessarily distinct). Suppose

$$f(t) = a_0 + a_1 t + \cdots + a_n t^n.$$

If we expand the first product and equate coefficients with the second expression we find that

$$a_n = k$$
$$a_{n-1} = -k(\alpha_1 + \cdots + \alpha_n)$$
$$a_{n-2} = k(\alpha_1\alpha_2 + \alpha_1\alpha_3 + \cdots + \alpha_{n-1}\alpha_n)$$
$$\cdots$$
$$a_0 = k(-1)^n\alpha_1\alpha_2 \cdots \alpha_n.$$

The expressions in $\alpha_1, \cdots, \alpha_n$ on the right (neglecting factors $\pm k$) have a special name.

Definition. The rth *elementary symmetric polynomial*

$$s_r(t_1, \cdots, t_n)$$

in the indeterminates $t_1, \cdots, t_n$ is the sum of all possible distinct products, taken r at a time, of the elements $t_1, \cdots, t_n$.

By abuse of language the result of substituting given elements $\alpha_1, \cdots, \alpha_n$ for $t_1, \cdots, t_n$ is called the rth elementary symmetric polynomial in $\alpha_1, \cdots, \alpha_n$. The above equations can be summed up as

$$a_{n-r} = k(-1)^r s_r(\alpha_1, \cdots, \alpha_n).$$

These polynomials are symmetric in the sense that they are unchanged by permutations of the indeterminates t_i. There are other symmetric polynomials apart from the elementary ones, for example $t_1^2 + \cdots + t_n^2$; but they can all be expressed in terms of elementary symmetric polynomials. (A general formula for $t_1^m + \cdots + t_n^m$ was found by Isaac Newton.) We quote without proof:

2.9 Theorem. *Over a field K, any symmetric polynomial in $t_1, \cdots, t_n$ can be expressed as a polynomial of smaller or equal degree in the elementary symmetric polynomials $s_r(t_1, \cdots, t_n)$ $(r = 0, \cdots, n)$.*

A slightly weaker version of this result is proved in corollary 15.4. We need 2.9 to prove that π is transcendental (Chapter 6). The quickest proof of 2.9 is by induction, and can be found in any of the older algebra texts (e.g. Salmon [25] p. 57, Van der Waerden [5] p. 81). As an example of the theorem in action we cite the equation

$$t_1^2 + t_2^2 = (t_1 + t_2)^2 - 2t_1 t_2 = s_1^2 - 2s_2.$$

Exercises

2.1 Decide the irreducibility or otherwise of the following polynomials:
(a) $t^4 + 1$ over **R**.
(b) $t^4 + 1$ over **Q**.
(c) $t^7 + 11t^3 - 33t + 22$ over **Q**.
(d) $t^4 + t^3 + t^2 + t + 1$ over **Q**.
(e) $t^3 - 7t^2 + 3t + 3$ over **Q**.
(f) $t^4 + 7$ over $\mathbf{Z}_{17}$.
(g) $t^3 - 5$ over $\mathbf{Z}_{11}$.
(h) $t^2 - \alpha t + \beta$ over the field of exercise 1.6.

2.2 In each of the above cases factorize the polynomial into irreducibles.

2.3 If K is a field with infinitely many elements and f, g are polynomials over K such that $f(\alpha) = g(\alpha)$ for all $\alpha \in K$, prove that $f = g$. Can you weaken the hypotheses a little?

2.4 Let K be a field. Say that a polynomial f over K is *prime* if whenever $f \mid gh$ either $f \mid g$ or $f \mid h$. Show that a polynomial $f \neq 0$ is prime if and only if it is irreducible. (The analogous result fails in more general rings.)

2.5 Find the zeros of the following polynomials; first over **Q**, then **R**, then **C**.
(a) $t^3 + 1$.
(b) $t^3 - 6t^2 + 11t - 6$.

(c) $t^5 + t + 1$.
(d) $t^2 + 1$.
(e) $t^4 + t^3 + t^2 + t + 1$.
(f) $t^4 - 6t^2 + 11$.

2.6 Write down all possible polynomials of the form $t^2 + at + b$ over $\mathbf{Z}_5$. Find out which are irreducible. In each case work out $a^2 - 4b$. What do you notice? Can you prove it?

2.7 Give a criterion for the irreducibility of a quadratic (degree 2) polynomial over any field of characteristic $\neq 2$.

2.8 Show that for any prime p the field $\mathbf{Z}_p(t)$ has characteristic p. Are $\mathbf{Z}_p$ and $\mathbf{Z}_p(t)$ isomorphic?

2.9 Express in terms of elementary symmetric polynomials of α, β, γ the following:
(a) $\alpha^2 + \beta^2 + \gamma^2$.
(b) $\alpha^3 + \beta^3 + \gamma^3$.
(c) $\alpha^2\beta + \alpha^2\gamma + \beta^2\alpha + \beta^2\gamma + \gamma^2\alpha + \gamma^2\beta$.
(d) $\alpha^2\beta^2\gamma^2$.
(e) $(\alpha - \beta)^2 + (\beta - \gamma)^2 + (\gamma - \alpha)^2$.
(f) $(\alpha - \beta)^2(\beta - \gamma)^2(\gamma - \alpha)^2$.
(g) $\alpha^2 + \beta\gamma$.

2.10 Why is it not sensible to try to solve a polynomial equation by solving the equations for the symmetric polynomials of the zeros? (Hint: If in doubt, try it out – on a cubic.)

2.11 Mark the following true or false.
(a) Every polynomial over a field K has a zero in K.
(b) Every polynomial which is irreducible over $\mathbf{Q}$ is also irreducible over $\mathbf{R}$.
(c) Every polynomial irreducible over $\mathbf{Z}$ is also irreducible over $\mathbf{Q}$.

(d) Linear polynomials are irreducible.

(e) All symmetric polynomials are elementary.

(f) Any polynomial in elementary symmetric polynomials is itself symmetric.

(g) There are infinitely many irreducible polynomials over $\mathbf{Q}$.

(h) Polynomials which are coprime have different degrees.

(i) A polynomial of prime degree is irreducible.

(j) A polynomial of composite degree is reducible.

CHAPTER THREE

Field extensions

Galois's original theory was couched in terms of polynomials over the complex field. The modern approach is a consequence of the methods used, in the 1920's and 1930's, to generalize the theory to arbitrary fields. From this viewpoint the central object of study ceases to be a polynomial, and becomes instead a 'field extension' related to a polynomial. Every polynomial f over a field K defines another field L containing K (or at any rate a subfield isomorphic to K). There are considerable advantages in setting up the theory from this field-theoretic point of view, and introducing polynomials at a later stage.

In this chapter we shall define field extensions and explain the link with polynomials. We shall also classify certain basic types of extension and give methods by which these may be constructed.

Field extensions

Roughly speaking, a field L is an extension of another field K if K is a subfield of L. For technical reasons this definition is too restrictive; we wish to allow cases where L contains a subfield *isomorphic* to K, but not necessarily equal.

Definition. A *field extension* is a monomorphism $i: K \to L$

where K and L are fields.

K is the *small* field, L the *large* field.

Notice that with a strict set-theoretic definition of *function*, the map i determines both K and L.

Examples

1 The inclusion maps $i_1 : \mathbf{Q} \to \mathbf{R}$, $i_2 : \mathbf{R} \to \mathbf{C}$, and $i_3 : \mathbf{Q} \to \mathbf{C}$ are all field extensions.

2 If K is any field, and $K(t)$ is the field of rational expressions over K, there is a natural monomorphism $i : K \to K(t)$ mapping each element of K to the corresponding constant polynomial. This is again a field extension.

3 Let $\mathbf{P}$ be the set of real numbers of the form $p + q\sqrt{2}$ where $p, q \in \mathbf{Q}$. $\mathbf{P}$ is a subfield of $\mathbf{R}$, since

$$(p + q\sqrt{2})^{-1} = \frac{p}{p^2 - 2q^2} - \frac{q}{p^2 - 2q^2}\sqrt{2}$$

if p and q are non-zero. The inclusion map $i : \mathbf{Q} \to \mathbf{P}$ is a field extension.

If $i : K \to L$ is a field extension, we can usually identify K with its image $i(K)$, so that i can be thought of as an inclusion map and K can be thought of as a subfield of L. Under these circumstances we use the notation

$$L : K$$

for the extension, and say that L *is an extension of* K.

In future we shall make the identification of K and $i(K)$ whenever this is legitimate.

The next concept is one which pervades much of abstract algebra:

Definition. Let K be a field, X a non-empty subset of K. Then the subfield of K *generated by* X is the intersection of all subfields of K which contain X.

The reader should convince himself that this definition is equivalent to either of the following:

(a) The smallest subfield of K which contains X.

(b) The set of all elements of K which can be obtained from elements of X by a finite sequence of field operations.

Example. We shall find the subfield of **C** generated by $X = \{1, i\}$. (Whenever we are talking of **C** the symbol i will denote $\sqrt{-1}$, as usual.) Let L be this subfield. Then L must contain the prime subfield **Q** of **C**; and since L is closed under the field operations it must contain all elements of the form

$$p + qi$$

where $p, q \in$ **Q**. Let M be the set of all such elements. We claim that M is a field. Clearly M is closed under sums and products; further

$$(p + iq)^{-1} = \frac{p}{p^2 + q^2} - \frac{q}{p^2 + q^2}i$$

so that every non-zero element of M has a multiplicative inverse in M. Hence M is a field, and contains X. Since L is the smallest subfield containing X, we have $L \subseteq M$. But $M \subseteq L$ by definition. Hence $L = M$, and we have found a description of the subfield generated by X.

In the case of a field extension $L:K$ we are more interested in fields lying *between* K and L. This means that we can restrict our attention to subsets X which contain K; equivalently sets of the form $K \cup Y$ where $Y \subseteq L$.

Definition. If $L:K$ is an extension and Y is a subset of L, then the subfield of L generated by $K \cup Y$ is written

$$K(Y)$$

and is said to be obtained from K by *adjoining* Y.

Notice that $K(Y)$ is in general considerably larger than $K \cup Y$.

This notation is open to all sorts of useful abuses. If Y

has a single element y we write $K(y)$ instead of $K(\{y\})$; and
in the same spirit $K(y_1, \cdots, y_n)$ will replace $K(\{y_1, \cdots, y_n\})$.

Examples

1 The subfield $\mathbf{R}(i)$ of $\mathbf{C}$ must contain all elements $x + iy$
where $x, y \in \mathbf{R}$. So $\mathbf{C} = \mathbf{R}(i)$.

2 Let K be any field, $K(t)$ the field of rational expressions
in t over K. This notation would appear to be ambiguous,
in that $K(t)$ also denotes the subfield generated by $K \cup \{t\}$.
But this subfield, since it is closed under the field operations,
must contain all rational expressions in t; hence it is the
whole field of rational expressions. Thus $K(t)$ has the same
meaning under either interpretation.

3 The subfield of $\mathbf{R}$ consisting of all elements $p + q\sqrt{2}$ where
$p, q \in \mathbf{Q}$ is easily seen to be $\mathbf{Q}(\sqrt{2})$.

4 It is not always true that a field of the form $K(\alpha)$ consists
of all elements of the form $j + k\alpha$ where $j, k \in K$. It certainly
contains all such elements, but they need not form a field.
For example, in $\mathbf{R} : \mathbf{Q}$ let α be the real cube root of 2, and
consider $\mathbf{Q}(\alpha)$. This is in fact the set of all elements of $\mathbf{R}$ of
the form $p + q\alpha + r\alpha^2$, where $p, q, r \in \mathbf{Q}$. To show this, we
prove that the set of such elements is a field. The only diffi-
culty is finding a multiplicative inverse; the reader should
work out the details.

Simple extensions

We define a special kind of field extension.

Definition. A *simple extension* is an extension $L : K$ having
the property that $L = K(\alpha)$ for some $\alpha \in L$.

The examples discussed above are all simple extensions.
On the other hand $\mathbf{R} : \mathbf{Q}$ is not a simple extension (see
Exercise 3.6). Note that *a simple extension need not be
presented in a form which makes its simplicity obvious.* For
example, $\mathbf{Q}(\sqrt{2}, \sqrt{3}) : \mathbf{Q}$ is a simple extension, even though
two elements are adjoined, not just one. For consider the

field $\mathbf{Q}(\sqrt{2}+\sqrt{3})$. This contains the elements $(\sqrt{2}+\sqrt{3})^2$ $= 5+2\sqrt{6}$, hence $\sqrt{6}$; hence $\sqrt{6}(\sqrt{2}+\sqrt{3}) = 2\sqrt{3}+3\sqrt{2}$; hence $\sqrt{2}$ and $\sqrt{3}$. Thus $\mathbf{Q}(\sqrt{2}, \sqrt{3}) = \mathbf{Q}(\sqrt{2}+\sqrt{3})$.

Our aim for the rest of this chapter is to classify all possible simple extensions. We first formulate the concept of isomorphism of extensions, then develop techniques for constructing simple extensions, and finally show that up to isomorphism we have constructed all possible simple extensions.

Definition. An *isomorphism* between two field extensions $i:K \to K^*, j:L \to L^*$ is a pair (λ, μ) of field isomorphisms $\lambda:K \to L, \mu:K^* \to L^*$, such that for all $k \in K$

$$j(\lambda(k)) = \mu(i(k)).$$

Another, more pictorial, way of putting this is to say that the *diagram*

$$\begin{array}{ccc} K & \xrightarrow{i} & K^* \\ \lambda \downarrow & & \downarrow \mu \\ L & \xrightarrow[j]{} & L^* \end{array}$$

commutes – that is, the two paths from K to L^* give the same map.

The reason for making the definition this way is that as well as the field structure being preserved by isomorphism, the embedding of the small field in the large one is also preserved.

Various identifications may be made. If we identify K and $i(K)$, and L and $j(L)$, then i and j are inclusions, and the commutativity condition now becomes

$$\mu|_K = \lambda$$

where $\mu|_K$ denotes the restriction of μ to K. If we further identify K and L then λ becomes the identity, and so $\mu|_K$ is the identity. In what follows we shall attempt to use these 'identified' conditions wherever possible. But on a few occasions (notably Theorem 8.3) we shall need the full generality of the first definition.

Constructing simple extensions

Any attempt to develop a theory of simple extensions leads rapidly to a fundamental dichotomy.

Definition. Let $K(\alpha):K$ be a simple extension. If there exists a non-zero polynomial p over K such that $p(\alpha) = 0$ then α is an *algebraic* element over K and the extension is a *simple algebraic extension*. Otherwise α is *transcendental* over K and $K(\alpha):K$ is a *simple transcendental extension*.

The next result gives a way of constructing a simple transcendental extension of any field.

3.1 Theorem. *The field of rational expressions $K(t)$ is a simple transcendental extension of the field K.*

Proof. Clearly it is a simple extension. If p is a polynomial over K such that $p(t) = 0$ then $p = 0$ by definition of $K(t)$.

The construction of simple algebraic extensions is a much more delicate problem. First we need a technical definition.

Definition. A polynomial

$$f(t) = a_0 + a_1 t + \cdots + a_n t^n$$

over a field K is *monic* if $a_n = 1$.

Clearly every polynomial is a constant multiple of some monic polynomial; and for a non-zero polynomial this monic polynomial is unique. Further, the product of two monic polynomials is again monic.

Now suppose that $K(\alpha):K$ is a simple algebraic extension. There is a polynomial p over K such that $p(\alpha) = 0$. We may suppose that p is monic. There is at least one monic polynomial of smallest degree of which α is a zero. If p, q are two such, then $p(\alpha) - q(\alpha) = 0$. If $p \neq q$ some constant multiple of $p - q$ is a monic polynomial of which α is a zero, contrary to the definition. Hence there is a unique monic polynomial p of smallest degree such that $p(\alpha) = 0$.

Definition. Let $L:K$ be a field extension, and suppose that $\alpha \in L$ is algebraic over K. Then the *minimum polynomial* of α over K is the unique monic polynomial m over K of smallest degree such that $m(\alpha) = 0$.

For example, $i \in \mathbf{C}$ is algebraic over $\mathbf{R}$. If we let $m(t) = t^2 + 1$ then $m(i) = 0$. Clearly m is monic. The only monic polynomials over $\mathbf{R}$ of smaller degree are those of the form $t + r$ ($r \in \mathbf{R}$) or 1. But i cannot be a zero of any of these, or else we would have $i \in \mathbf{R}$. Hence the minimum polynomial of i over $\mathbf{R}$ is $t^2 + 1$.

It is natural to ask which polynomials can occur as minimum polynomials. The next lemma provides information on this question.

3.2 Lemma. *If α is an algebraic element over the field K, then the minimum polynomial of α over K is irreducible over K. It divides every polynomial of which α is a zero.*

Proof. Suppose that the minimum polynomial m of α over K is reducible, so that $m = fg$ where f and g are of smaller degree. We may assume f and g are monic. Since $m(\alpha) = 0$ we have $f(\alpha)g(\alpha) = 0$ so either $f(\alpha) = 0$ or $g(\alpha) = 0$. But this contradicts the definition of m. Hence m is irreducible over K.

Now suppose that p is a polynomial over K such that $p(\alpha) = 0$. By the division algorithm there exist polynomials q and r over K such that $p = mq + r$ and $\partial r < \partial m$. Then $0 = p(\alpha) = 0 + r(\alpha)$. If $r \neq 0$ a suitable constant multiple of it is monic, which contradicts the definition of m. Therefore $r = 0$ so that p divides m.

Remark for the sophisticate. Using higher-powered machinery we can express ourselves more succinctly. The ring $K[t]$ is a principal ideal domain, and $\{p : p(\alpha) = 0\}$ is an ideal, so a principal ideal. Its unique monic generator is m. For details, consult Hartley and Hawkes [14] pp. 59 ff.

Given any field K and any irreducible monic polynomial m over K we shall construct an extension $K(\alpha):K$ such that α has minimum polynomial m over K. We need two preliminary lemmas.

3.3 Lemma. *If ϕ is a ring homomorphism from a field K into a ring R and $\phi \neq 0$ then ϕ is a monomorphism.*

Proof. The kernel of ϕ is an ideal of K. But K, being a field, has no ideals other than 0 and K itself. Since $\phi \neq 0$ the kernel is not K, so must be 0. Therefore ϕ is a monomorphism.

Note that this lemma fails if K is just a ring: the natural map $\mathbf{Z} \to \mathbf{Z}_2$ is neither zero nor a monomorphism.

3.4 Lemma. *If m is an irreducible polynomial over the field K, and I is the ideal of $K[t]$ consisting of all multiples of m, then the quotient ring $K[t]/I$ is a field.*

Proof. Let the coset $I+f$ be a nonzero element of $S = K[t]/I$. Since m is irreducible m and f are coprime. Hence by 1.12 there exist polynomials a, b over K such that

$$af + bm = 1.$$

Then

$$(I+a)(I+f) + (I+b)(I+m) = I+1.$$

But $I+m = I$ is the zero element of S, and $I+1$ is the identity element, so that

$$(I+a)(I+f) = I+1$$

and $I+f$ has inverse $I+a$. Hence S is a field.

We now have what we need to prove:

3.5 Theorem. *If K is any field and m is any irreducible monic polynomial over K, then there exists an extension $K(\alpha):K$ such that α has minimum polynomial m over K.*

Proof. There is a natural monomorphism $i:K \to K[t]$. Let I be the ideal of $K[t]$ consisting of all multiples of m, let $S = K[t]/I$, and let v be the natural homomorphism $K[t] \to S$. By Lemma 3.4 S is a field, and by Lemma 3.3 the composite map vi is a monomorphism. Identify K with its image $v(i(K))$, and let $\alpha = I+t$. Clearly $S = K(\alpha)$. Since $m \in I$ we have $m(\alpha) = I$, and I is the zero element of S. Since m is irreducible and monic it must be the minimum polynomial of α. For if p is the minimum polynomial, then by 3.2 $p|m$. Therefore $p = m$.

There is another way to look at this construction. From each coset $I+f$ we can select a unique polynomial of degree smaller than ∂m. The field structure on S induces operations on the set of representative polynomials as follows: addition is as usual. Multiplication is as usual except that after multiplying it is necessary to take the remainder on division by m. It is possible to *define* $K(\alpha)$ in this way.

Classifying simple extensions

We shall now demonstrate that the above methods suffice for the construction of all possible simple extensions (up to isomorphism). Again transcendental extensions are easily dealt with.

3.6 Theorem. *Every simple transcendental extension $K(\alpha):K$ is isomorphic to the extension $K(t):K$ of rational expressions in t over K. The isomorphism can be chosen to carry t into α.*

Proof. Define a map $\phi:K(t) \to K(\alpha)$ by

$$\phi(f(t)/g(t)) = f(\alpha)/g(\alpha).$$

If $g \neq 0$ then $g(\alpha) \neq 0$ (since α is transcendental) so this definition makes sense. It is clearly a homomorphism, and is a monomorphism by 3.3. It is clearly onto, and so is an isomorphism. Further, $\phi|_K$ is the identity, so that ϕ defines an isomorphism of extensions. And $\phi(t) = \alpha$.

To deal with algebraic extensions we must establish a standard form for elements of the large field.

3.7 Lemma. *Let $K(\alpha):K$ be a simple algebraic extension, where α has minimum polynomial m over K. Then any element of $K(\alpha)$ has a unique expression in the form $p(\alpha)$ where p is a polynomial over K and $\partial p < \partial m$.*

Proof. Every element of $K(\alpha)$ can be expressed in the form $f(\alpha)/g(\alpha)$ where $f, g \in K[t]$ and $g(\alpha) \neq 0$ (since the set of all such elements is a field, contains K and α, and lies inside $K(\alpha)$). Since $g(\alpha) \neq 0$, m does not divide g; and since m is irreducible m and g are coprime. By 1.12 there exist polynomials a, b over K such that $ag + bm = 1$. Hence $a(\alpha)g(\alpha) = 1$, so that $f(\alpha)/g(\alpha) = f(\alpha)a(\alpha) = h(\alpha)$ for some polynomial h over K. Let r be the remainder on dividing h by m. Then $r(\alpha) = h(\alpha)$. Since $\partial r < \partial m$, the existence of such an expression is proved.

Now we show uniqueness. Suppose that $f(\alpha) = g(\alpha)$ where $\partial f, \partial g < \partial m$. If $e = f - g$ then $e(\alpha) = 0$ and $\partial e < \partial m$. By definition of m we have $e = 0$, so that $f = g$. The lemma is proved.

Example. Suppose $K = \mathbf{R}$, $m(t) = t^2 + t + 1$. According to the above lemma, if α has minimum polynomial m over K then every element of $K(\alpha)$ is a polynomial in α of degree <2. Consider the element $(3\alpha^2 + 2)/(\alpha + 4)$. Observe that

$$1 = \tfrac{1}{13}(t^2 + t + 1) - \frac{t-3}{13}(t+4)$$

so that

$$1/(\alpha + 4) = -(\alpha - 3)/13.$$

Therefore

$$(3\alpha^2 + 2)/(\alpha + 4) = \tfrac{1}{13}(3\alpha^2 + 2)(\alpha - 3)$$

$$- \tfrac{1}{13} 3(-\alpha - 1) + 2)(\alpha - 3)$$

$$- \tfrac{1}{13}(-3\alpha^2 + 8\alpha + 3)$$

$$- \tfrac{1}{13}(11\alpha + 6)$$

$$- \tfrac{11}{13}\alpha - \tfrac{6}{13}.$$

We can now prove a preliminary version of the result that K and m between them determine the extension $K(\alpha)$.

3.8 Theorem. *Suppose $K(\alpha):K$ and $K(\beta):K$ are simple algebraic extensions, such that α and β have the same minimum polynomial m over K. Then the two extensions are isomorphic, and the isomorphism of the large fields can be taken to map α to β.*

Proof. By 3.7 every element $x \in K(\alpha)$ is uniquely expressible in the form

$$x = x_0 + x_1\alpha + \cdots + x_n\alpha^n \quad (x_1, \cdots, x_n \in K)$$

where $n = \partial m - 1$. Define a map $\phi : K(\alpha) \to K(\beta)$ by

$$\phi(x) = x_0 + x_1\beta + \cdots + x_n\beta^n.$$

By 3.7 ϕ is onto and $1-1$. Obviously $\phi(x+y) = \phi(x) + \phi(y)$. We shall show that $\phi(xy) = \phi(x)\phi(y)$, for any $x, y \in K(\alpha)$. Let $x = f(\alpha)$, $y = g(\alpha)$, $xy = h(\alpha)$; where f, g, h are polynomials over K of degree $< \partial m$. Then

$$f(\alpha)g(\alpha) - h(\alpha) = xy - xy = 0.$$

By 3.2 m divides $fg - h$, so there exists a polynomial q over K such that $fg = mq + h$. Since $\partial h < \partial m$ it follows that h is the remainder on dividing fg by m. By the same reasoning we must have $f(\beta)g(\beta) = h(\beta)$. Thus $\phi(xy) = h(\beta) = f(\beta)g(\beta) = \phi(x)\phi(y)$, so ϕ is an isomorphism. Since ϕ is the identity on K the two extensions are isomorphic. And clearly $\phi(\alpha) = \beta$.

For certain later applications we need a slightly stronger version of the previous theorem, to cover extensions of isomorphic (rather than identical) fields. Before we can state the more general theorem we need the following:

Definition. Let $i:K \rightarrow L$ be a field monomorphism. There is a monomorphism $\hat{\imath}:K[t] \rightarrow L[t]$ defined by

$$\hat{\imath}(k_0 + k_1 t + \cdots + k_n t^n) = i(k_0) + i(k_1)t + \cdots + i(k_n)t^n$$

($k_0, \cdots, k_n \in K$). If i is an isomorphism so is $\hat{\imath}$.

The hat is unnecessary and may be dispensed with; and in future we shall use the same symbol i for the map between fields and for its extension to the polynomial rings. This will not cause confusion since $\hat{\imath}(k) = i(k)$ for any $k \in K$.

3.9 Theorem. *Suppose K and L are fields and $i:K \rightarrow L$ is an isomorphism. Let $K(\alpha)$, $L(\beta)$ be simple algebraic extensions of K and L respectively, such that α has minimum polynomial $m_\alpha(t)$ over K and β has minimum polynomial $m_\beta(t)$ over L. Suppose further that $m_\beta(t) = i(m_\alpha(t))$. Then there exists an isomorphism $j:K(\alpha) \rightarrow L(\beta)$ such that $j|_K = i$ and $j(\alpha) = \beta$.*

Proof. We have the diagram

$$\begin{array}{ccc} K & \rightarrow & K(\alpha) \\ i \downarrow & & \downarrow j \\ L & \rightarrow & L(\beta) \end{array}$$

(where the dotted arrow indicates that j has not yet been found). Using the proof of 3.8 as a guide, define j as follows: every element of $K(\alpha)$ is of the form $p(\alpha)$ for a polynomial p over K of degree $< \partial m_\alpha$. Let $j(p(\alpha)) = (i(p))(\beta)$ where $i(p)$ is defined as above. The detailed proof follows the lines of 3.8 and is left to the reader.

The point of this theorem is that the given map i can be extended to a map j between the larger fields. Such *extension theorems*, saying that under suitable conditions maps between sub-objects can be extended to maps between objects, constitute important weapons in the mathematician's armoury. Using them we can extend our knowledge from small structures to large ones in a sequence of simple steps. Not surprisingly, extension theorems are usually hard to come by!

Theorem 3.9 implies that under the given hypotheses the *extensions* $K(\alpha):K$ and $L(\beta):L$ are isomorphic. This allows us to identify K with L *and* $K(\alpha)$ with $L(\beta)$, via the maps i and j.

Theorems 3.5 and 3.8 together give a complete characterization of simple algebraic extensions in terms of polynomials. To each extension corresponds an irreducible monic polynomial; and given the small field and this polynomial we can reconstruct the extension.

Note that the correspondence is not strictly 1–1: isomorphic extensions may give different polynomials, since there is latitude of choice for α. No difficulties arise because of this.

Exercises

3.1 Prove that isomorphism of field extensions is an equivalence relation.

3.2 Find the subfields of **C** generated by:
(a) $\{0, 1\}$
(b) $\{0\}$
(c) $\{0, 1, i\}$
(d) $\{i, \sqrt{2}\}$
(e) $\{\sqrt{2}, \sqrt{3}\}$
(f) **R**
(g) **R** $\cup \{i\}$.

3.3 Describe the subfields of **C** of the form
(a) $\mathbf{Q}(\sqrt{2})$
(b) $\mathbf{Q}(i)$
(c) $\mathbf{Q}(\alpha)$ where α is the real cube root of 2

(d) $\mathbf{Q}(\sqrt{5}, \sqrt{7})$

(e) $\mathbf{Q}(i\sqrt{11})$.

3.4 Let $K = \mathbf{Z}_2$. Describe the subfields of $K(t)$ of the form:

(a) $K(t^2)$

(b) $K(t+1)$

(c) $K(t^5)$

(d) $K(t^2+1)$.

3.5 Of the extensions defined in Exercises 3.3 and 3.4, which are simple algebraic? Which are simple transcendental?

3.6 Show that $\mathbf{R}$ is not a simple extension of $\mathbf{Q}$ as follows:

(a) $\mathbf{Q}$ is countable.

(b) Any simple extension of a countable field is countable.

(c) $\mathbf{R}$ is not countable.

3.7 Prove a transcendental version of Theorem 3.9, modelled on Theorems 3.9 and 3.1.

3.8 Find the minimum polynomials over the small field of the following elements in the following extensions:

(a) i in $\mathbf{C}:\mathbf{Q}$

(b) i in $\mathbf{C}:\mathbf{R}$

(c) $\sqrt{2}$ in $\mathbf{R}:\mathbf{Q}$

(d) $(\sqrt{5}+1)/2$ in $\mathbf{C}:\mathbf{Q}$

(e) $(i\sqrt{3}-1)/2$ in $\mathbf{C}:\mathbf{Q}$

(f) α in $K:P$ where K is the field of Exercise 1.6 and P is its prime subfield.

(g) α in $\mathbf{Z}_3(t)(\alpha):\mathbf{Z}_3(t)$ where t is indeterminate and $\alpha^2 = t+1$.

3.9 Show that if α has minimum polynomial t^2-2 over $\mathbf{Q}$ and β has minimum polynomial t^2-4t+2 over

$\mathbf{Q}$, then the extensions $\mathbf{Q}(\alpha):\mathbf{Q}$ and $\mathbf{Q}(\beta):\mathbf{Q}$ are isomorphic.

3.10 For which of the following values of $m(t)$ do there exist extensions $K(\alpha)$ of K for which α has minimum polynomial $m(t)$?
(a) $m(t) = t^2 - 4$, $K = \mathbf{R}$
(b) $m(t) = t^2 + 1$, $K = \mathbf{Z}_3$
(c) $m(t) = t^2 + 1$, $K = \mathbf{Z}_5$
(d) $m(t) = t^7 - 3t^6 + 4t^3 - t - 1$, $K = \mathbf{R}$.

3.11 Let K be any field of characteristic $\neq 2$ and $m(t)$ a quadratic polynomial over K (i.e. $\partial m = 2$). Show that $m(t)$ has both zeros in an extension field $K(\alpha)$ of K where $\alpha^2 = k \in K$. Thus allowing 'square roots' $\sqrt{k}$ enables us to solve all quadratic equations over K.

3.12 Show that for fields of characteristic 2 there exist quadratic equations which cannot be solved by adjoining square roots of elements in the field. (Hint: try $\mathbf{Z}_2$).

3.13 Show that we *can* solve quadratic equations over a field of characteristic 2 if as well as square roots we adjoin elements $\sqrt[*]{k}$ defined to be solutions of the equation

$$(\sqrt[*]{k})^2 + \sqrt[*]{k} = k.$$

3.14 Show that the two zeros of $t^2 + t - k = 0$ in the previous question are $\sqrt[*]{k}$ and $1 + \sqrt[*]{k}$.

3.15 Let $K = \mathbf{Z}_3$. Find all irreducible quadratics over K, and construct all possible extensions of K by an element with quadratic minimum polynomial. Into how many isomorphism classes do these extensions fall? How many elements do they have?

3.16 Construct extensions $\mathbf{Q}(\alpha):\mathbf{Q}$ where α has the following minimum polynomials over $\mathbf{Q}$:
(a) $t^2 - 5$
(b) $t^4 + t^3 + t^2 + t + 1$
(c) $t^3 + 2$.

3.17 Find a field with 8 elements.

3.18 Is $\mathbf{Q}(\sqrt{2}, \sqrt{3}, \sqrt{5}):\mathbf{Q}$ a simple extension?

3.19 Suppose $m(t)$ is irreducible over K, and α has minimum polynomial $m(t)$ over K. Does $m(t)$ necessarily factorize over $K(\alpha)$ into linear (degree 1) polynomials? Hint: try $K = \mathbf{Q}$, $\alpha =$ the real cube root of 2.

3.20 Mark the following true or false.
(a) Every field has non-trivial extensions.
(b) Every field has non-trivial algebraic extensions.
(c) Every simple extension is algebraic.
(d) Every extension is simple.
(e) All simple algebraic extensions are isomorphic.
(f) All simple transcendental extensions of a given field are isomorphic.
(g) Every minimum polynomial is monic.
(h) Monic polynomials are always irreducible.
(i) Every polynomial is a constant multiple of an irreducible polynomial.
(j) It is always safe to identify isomorphic fields.

CHAPTER FOUR

The degree of an extension

A technique which has become very useful in mathematics is that of associating with a given structure a different one, of a type better understood. This method became rampant in algebraic topology, so that topologists were forced to formalize these methods; and it then became clear that they underlay a great deal of mathematics.

In this chapter we exploit the technique by associating with any field extension a vector space. This places at our disposal the machinery of linear algebra – a very successful algebraic theory – and with its aid we can make considerable progress. The machinery is sufficiently powerful to solve three notorious problems which remained unanswered for over 2000 years. We shall discuss these problems in the next chapter, and devote the present chapter to developing the theory.

It is not hard to define a vector space structure on a field extension. It already has one! More precisely:

4.1 Theorem. *If $L:K$ is a field extension, the operations*

$$(\lambda, u) \to \lambda u \qquad (\lambda \in K, u \in L)$$

$$(u, v) \to u + v \qquad (u, v \in L)$$

define on L the structure of a vector space over K.

Proof. Each axiom for a vector space is either a field axiom or a restricted case of a field axiom.

The operation which turns $L:K$ into a vector space is an example of what is called a 'forgetful functor': it simply forgets some of the structure. The resulting coarsening of our perceptions allows us to isolate aspects of the structure which would otherwise be lost in a mass of irrelevant detail.

A vector space over a given field is uniquely determined (up to isomorphism) by its dimension. The following definition is the traditional terminology in the context of field extensions:

Definition. The *degree* $[L:K]$ of a field extension $L:K$ is the dimension of L considered as a vector space over K.

Examples

1 The complex numbers **C** are 2-dimensional over the real numbers **R**, $\{1, i\}$ being a basis. Hence $[\mathbf{C}:\mathbf{R}] = 2$.
2 $[\mathbf{C}(t):\mathbf{C}]$ is infinite, since the elements $1, t, t^2, \cdots$ are linearly independent over **C**.
3 Let K be the field of 4 elements defined in Exercise 1.6 and let P be its prime subfield, which is isomorphic to $\mathbf{Z}_2$. The elements $\{1, \alpha\}$ form a basis for K over P, so that $[K:P] = 2$.

Isomorphic field extensions obviously have the same degree.

The next theorem allows us to calculate the degree of a complicated extension if we know the degrees of certain simpler ones.

4.2 Theorem. If K, L, M are fields and $K \subseteq L \subseteq M$, then

$$[M:K] = [M:L][L:K].$$

Note: For those who are happy with infinite cardinals this formula needs no extra explanation; the product on the

right is just multiplication of cardinals. For those who are
not, the formula needs interpretation if any of the degrees
involved is infinite. This interpretation is the obvious one:
if either $[M:L]$ or $[L:K] = \infty$ then $[M:K] = \infty$; and if
$[M:K] = \infty$ then either $[M:L] = \infty$ or $[L:K] = \infty$.

Proof. Let $(x_i)_{i \in I}$ be a basis for L as vector space over K and
let $(y_j)_{j \in J}$ be a basis for M over L. For all $i \in I$ and $j \in J$ we
have $x_i \in L$, $y_j \in M$. We shall show that $(x_i y_j)_{i \in I, j \in J}$ is a basis
for M over K ($x_i y_j$ is the product in the field M). Since
dimensions are cardinalities of bases the theorem will
follow.

First we show linear independence. Suppose that some
finite linear combination of the putative basis elements is
zero, i.e.

$$\sum_{i,j} k_{ij} x_i y_j = 0 \quad (k_{ij} \in K).$$

We can rearrange this as

$$\sum_j (\sum_i k_{ij} x_i) y_j = 0.$$

Since the coefficients $\sum_i k_{ij} x_i$ lie in L and the y_j are linearly
independent over L we must have

$$\sum_i k_{ij} x_i = 0.$$

Repeating the argument inside L we find that $k_{ij} = 0$ for
all $i \in I, j \in J$. So the elements $x_i y_j$ are linearly independent
over K.

Finally we show that the $x_i y_j$ span M over K. Now any
element $x \in M$ can be written

$$x = \sum_j \lambda_j y_j$$

for suitable $\lambda_j \in L$, since the y_j span M over L. Similarly for
any $j \in J$

$$\lambda_j = \sum_i \lambda_{ij} x_i$$

for $\lambda_{ij} \in K$. Putting the pieces together we find that

$$x = \sum_{i,j} \lambda_{ij} x_i y_j$$

as required.

Example. Suppose we wish to find $[\mathbf{Q}(\sqrt{2}, \sqrt{3}):\mathbf{Q}]$. It is easy to see that $\{1, \sqrt{2}\}$ is a basis for $\mathbf{Q}(\sqrt{2})$ over $\mathbf{Q}$. It is a little harder to see that $\{1, \sqrt{3}\}$ is a basis for $\mathbf{Q}(\sqrt{2}, \sqrt{3})$ over $\mathbf{Q}(\sqrt{2})$. Hence

$$[\mathbf{Q}(\sqrt{2}, \sqrt{3}):\mathbf{Q}] = [\mathbf{Q}(\sqrt{2}, \sqrt{3}):\mathbf{Q}(\sqrt{2})][\mathbf{Q}(\sqrt{2}):\mathbf{Q}]$$

$$= 2.2 = 4.$$

The theorem even furnishes a basis, namely $\{1, \sqrt{2}, \sqrt{3}, \sqrt{6}\}$.

In the case of simple extensions we can easily find the degree.

4.3 Proposition. *Let $K(\alpha):K$ be a simple extension. If it is transcendental then $[K(\alpha):K] = \infty$. If it is algebraic then $[K(\alpha):K] = \partial m$ where m is the minimum polynomial of α over K.*

Proof. For the transcendental case it suffices to note that the elements $1, \alpha, \alpha^2, \cdots$ are linearly independent over K. For the algebraic case, we exhibit a basis. Let $\partial m = n$ and consider the elements $1, \alpha, \cdots \alpha^{n-1}$. By 3.7 they span $K(\alpha)$ over K, and by the uniqueness clause of 3.7 they are linearly independent. Therefore they form a basis, and $[K(\alpha):K] = n = \partial m$.

This result can be illustrated by a simple example. We know that $\mathbf{C} = \mathbf{R}(i)$ where i has minimum polynomial $t^2 + 1$, of degree 2. Hence $[\mathbf{C}:\mathbf{R}] = 2$, which agrees with our previous remarks.

Linear algebra is at its most powerful when dealing with finite-dimensional vector spaces. Accordingly we shall concentrate on field extensions which give rise to such vector spaces.

Definition. A *finite extension* is one whose degree is finite.

Proposition 4.3 implies that any simple algebraic extension is finite. The converse is not true. In order to state what *is* true we need:

Definition. An extension $L:K$ is *algebraic* if every element of L is algebraic over K.

Algebraic extensions need not be finite – for example, the algebraic numbers considered below. But every finite extension is algebraic:

4.4 Lemma. $L:K$ *is a finite extension if and only if L is algebraic over K and there exist finitely many elements $\alpha_1, \cdots, \alpha_s \in L$ such that $L = K(\alpha_1, \cdots, \alpha_s)$.*

Proof. A simple induction using 4.2 and 4.3. shows that any algebraic extension $K(\alpha_1, \cdots, \alpha_s):K$ is finite. Conversely, let $L:K$ be a finite extension. Then there is a basis $\{\alpha_1, \cdots, \alpha_s\}$ for L over K, whence $L = K(\alpha_1, \cdots, \alpha_s)$. It remains to show that $L:K$ is algebraic. Let x be any element of L and let $n = [L:K]$. The set $\{1, x, \cdots, x^n\}$ contains $n+1$ elements, which must therefore be linearly dependent over K. Hence

$$k_0 + k_1 x + \cdots + k_n x^n = 0$$

for $k_0, \cdots, k_n \in K$, and x is algebraic over K.

Algebraic numbers

Let **A** denote the set of all complex numbers which are algebraic over **Q**. The elements of **A** are called *algebraic numbers*. We shall use the techniques of this chapter to show that **A** is a field.

By 4.4 a complex number α lies in **A** if and only if $[\mathbf{Q}(\alpha):\mathbf{Q}] < \infty$. Let α, β be any two elements of **A**. Then $[\mathbf{Q}(\alpha, \beta):\mathbf{Q}] = [\mathbf{Q}(\alpha, \beta):\mathbf{Q}(\alpha)][\mathbf{Q}(\alpha):\mathbf{Q}] < \infty$. Therefore $[\mathbf{Q}(\alpha+\beta):\mathbf{Q}] < \infty$, $[\mathbf{Q}(-\alpha):\mathbf{Q}] < \infty$, $[\mathbf{Q}(\alpha\beta):\mathbf{Q}] < \infty$, and if $\alpha \neq 0$ then $[\ \ (\alpha^{-1}):\ \ ] < \infty$. Hence $\alpha + \beta$, $\alpha - \beta$, $\alpha\beta$, α^{-1} all lie in $\mathbf{A}$, and $\mathbf{A}$ is a field.

Note that $\mathbf{A}$ is obviously an algebraic extension of $\mathbf{Q}$. But it can be shown (see Exercise 4.8) that $\mathbf{A}$ has infinite degree over $\mathbf{Q}$. Hence not every algebraic extension is finite.

Exercises

4.1 Find the degrees of the following extensions:
   (a) $\mathbf{C}:\mathbf{Q}$
   (b) $\mathbf{Z}_5(t):\mathbf{Z}_5$
   (c) $\mathbf{R}(\sqrt{5}):\mathbf{R}$
   (d) $\mathbf{Q}(\alpha):\mathbf{Q}$ where α is the real cube root of 2
   (e) $\mathbf{Q}(\sqrt{3}, \sqrt{5}, \sqrt{11}):\mathbf{Q}$
   (f) $\mathbf{Q}(\sqrt{6}):\mathbf{Q}$
   (g) $\mathbf{Q}(\alpha):\mathbf{Q}$ where $\alpha^7 = 3$.

4.2 Show that every element of $\mathbf{Q}(\sqrt{5}, \sqrt{7})$ can be expressed uniquely in the form

$$p + q\sqrt{5} + r\sqrt{7} + s\sqrt{35}$$

where $p, q, r, s \in \mathbf{Q}$. Calculate explicitly the inverse of such an element.

4.3 If $[L:K]$ is a prime number show that the only fields M such that $K \subseteq M \subseteq L$ are K and L themselves.

4.4 If $[L:K] = 1$ show that $K = L$.

4.5 If $K = K_0 \subseteq K_1 \subseteq \cdots \subseteq K_r = L$ are fields, show that

$$[L:K] = [K_r:K_{r-1}] \cdots [K_2:K_1][K_1:K_0].$$

4.6 Prove that $[L:K]$ is finite if and only if $L = K(\alpha_1, \cdots, \alpha_r)$ where r is finite and each α_i is algebraic over K.

4.7 Starting from the fact that $\mathbf{R}$ is a vector space over $\mathbf{Q}$ show that there are functions $f:\mathbf{R} \to \mathbf{R}$ such that

$f(x+y) = f(x)+f(y)$ for all $x, y \in \mathbf{R}$, but such that $f(x)$ is not a constant multiple of x. Do such functions exist if we also require them to be continuous?

4.8 Let **A** be the field of algebraic numbers. Prove that $[\mathbf{A}:\mathbf{Q}] = \infty$ by using Eisenstein's Criterion to show that there exist irreducible polynomials over **Q** of arbitrarily large degree.

4.9 Assuming that every polynomial over **C** has a zero in **C**, prove that every polynomial over **A** has a zero in **A**.

4.10 Using 4.9 show that every algebraic extension of **A** is just **A** itself.

4.11 Let $L:K$ be an extension. Show that multiplication by a fixed element of L is a linear transformation of L considered as a vector space over K. When is this linear transformation non-singular?

4.12 Let $L:K$ be a finite extension and let p be an irreducible polynomial over K. Show that if ∂p and $[L:K]$ are coprime then p has no zeros in L.

4.13 If $L:K$ is algebraic and $M:L$ is algebraic is $M:K$ algebraic? Note that you may *not* assume the extensions are finite.

4.14 Prove that $\mathbf{Q}(\sqrt{3}, \sqrt{5}) = \mathbf{Q}(\sqrt{3}+\sqrt{5})$. Try to generalize your result.

4.15 Prove that the square roots of all prime numbers are linearly independent over **Q**.

4.16 Find a basis for $\mathbf{Q}(\sqrt{(1+\sqrt{3})})$ over **Q** and hence find the degree of $\mathbf{Q}(\sqrt{(1+\sqrt{3})}):\mathbf{Q}$. Now find the degree by a different method.

4.17 If $[L:K]$ is prime show that L is a simple extension of K.

4.18 Mark the following true or false.
   (a) Extensions of the same degree are isomorphic.
   (b) Isomorphic extensions have the same degree.
   (c) Every algebraic extension is finite.
   (d) Every transcendental extension is not finite.
   (e) Every element of **C** is algebraic over **R**.
   (f) Every extension of **R** is finite.
   (g) Every algebraic extension of **Q** is finite.
   (h) **A** is the largest subfield of **C** which is algebraic over **Q**.
   (i) Every vector space is isomorphic to the vector space corresponding to some field extension.
   (j) Every extension of a finite field is finite.

CHAPTER FIVE

Ruler and compasses

According to Plato the only 'perfect' geometrical figures are the straight line and the circle. In ancient Greek geometry this belief had the effect of restricting the instruments available for performing geometrical constructions to two: the ruler and the compasses. The ruler, furthermore, was a single unmarked straight edge.

With these instruments alone it is possible to perform a wide range of constructions. Lines can be divided into arbitrarily many equal segments, angles can be bisected, parallel lines drawn. Given any polygon it is possible to construct a square of equal area, or twice the area. And so on. However, there are many geometrical concepts which *intuitively* ought to be constructible, for which the tools of ruler and compasses are inadequate. There are three famous constructions which the Greeks could not perform: *Duplication of the cube*, *Trisection of the angle*, and *Quadrature of the circle*. These ask respectively for a cube twice the volume of a given cube, an angle one third the size of a given angle, and a square of area equal to a given circle.

It is not surprising that the Greeks found these constructions so difficult. They are impossible. But the Greeks had neither the methods to prove the impossibility nor, it appears, the suspicion that solutions did not exist. In consequence they expended considerable ingenuity in a fruitless search for solutions.

By going outside the Platonic constraints all three problems can be solved, and the Greeks found several constructions involving conic sections or more recondite curves such as the conchoid of Nichomedes or the quadratrix (see Klein [42], Coolidge [35]). Archimedes, tackling the problem of quadrature of the circle in a characteristically ingenious manner, *proved* a result which would now be written

$$3\tfrac{10}{71} < \pi < 3\tfrac{1}{7},$$

a remarkable achievement with the limited techniques available.

With the machinery now at our disposal it is relatively simple to give a complete answer to all three problems. We use coordinate geometry to express them in algebraic terms, and apply our theory of field extensions to the algebraic problems which arise.

Algebraic formulation

Our first step is to formalize the intuitive idea of a ruler-and-compass construction. Assume that a set P_0 of points in the Euclidean plane $\mathbf{R}^2$ is given, and consider operations of the following two kinds:

Operation 1 (Ruler): Through any two points of P_0 draw a straight line.

Operation 2 (Compasses): Draw a circle, whose centre is a point of P_0, and whose radius is equal to the distance between some pair of points in P_0.

(The Greeks preferred a restricted version of operation 2, namely: draw a circle, centre some point of P_0, and passing through some other point of P_0. Our operation 2 can be performed by a sequence of such operations, so it ultimately makes no difference which version we use. The one given above is more convenient for our purposes.)

Definition. The points of intersection of any two distinct lines or circles, drawn using operations 1 or 2, are said to be *constructible in one step* from P_0.

A point $r \in \mathbf{R}^2$ is *constructible from* P_0 if there is a finite sequence

$$r_1, \cdots, r_n = r$$

of points of $\mathbf{R}^2$ such that for each $i = 1, \cdots, n$ the point r_i is constructible in one step from the set

$$P_0 \cup \{r_1, \cdots, r_{i-1}\}.$$

Example. We shall show how the standard construction of the mid-point of a given line can be realized within our formal framework. Suppose we are given two points $p_1, p_2 \in \mathbf{R}^2$ (Fig. 2).

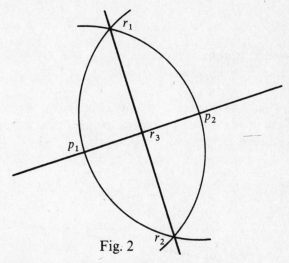

Fig. 2

Let $P_0 = \{p_1, p_2\}$.
(1) Draw the line $p_1 p_2$ (operation 1).
(2) Draw the circle centre p_1 of radius $p_1 p_2$ (operation 2).
(3) Draw the circle centre p_2 of radius $p_1 p_2$ (operation 2).
(4) Let r_1 and r_2 be the points of intersection of these circles.
(5) Draw the line $r_1 r_2$ (operation 1).
(6) Let r_3 be the intersection of the lines $p_1 p_2$ and $r_1 r_2$.
Then the sequence r_1, r_2, r_3 defines a construction of the mid-point of the line $p_1 p_2$.

Since a line is always specified by two points lying on it, and a circle by its centre and a point on its circumference, all the traditional geometrical constructions of Euclidean geometry fall within the scope of our formal definition.

Field theory enters in a fairly natural way. With each stage in the construction we associate the subfield of **R** generated by the coordinates of the points constructed. Thus let K_0 be the subfield of **R** generated by the x- and y-coordinates of the point in P_0. If r_i has coordinates (x_i, y_i) then inductively we define K_i to be the field obtained from K_{i-1} by adjoining x_i and y_i, that is to say,

$$K_i = K_{i-1}(x_i, y_i).$$

Clearly

$$K_0 \subseteq K_1 \subseteq \cdots \subseteq K_n \subseteq \mathbf{R}.$$

5.1 Lemma. *With the above notation, x_i and y_i are zeros in K_i of quadratic polynomials over K_{i-1}.*

Proof. There are three cases to consider: line meets line, line meets circle, and circle meets circle. Each case is handled by coordinate geometry; as an example we shall do the case 'line meets circle'.

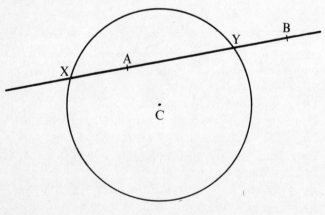

Fig. 3

Let A, B, C be points whose coordinates $(p, q), (r, s), (t, u)$ lie in K_{i-1}. Draw the line AB and the circle centre C, radius w, where $w^2 \in K_{i-1}$, as in Fig. 3. (Note that w^2 lies in K_{i-1} since w is the distance between two points whose coordinates are in K_{i-1}. Use Pythagoras.) The equation of the line AB is

$$\frac{x-p}{r-p} = \frac{y-q}{s-q} \tag{1}$$

and the equation of the circle is

$$(x-t)^2 + (y-u)^2 = w^2. \tag{2}$$

Solving (1) and (2) we obtain

$$(x-t)^2 + \left(\frac{(s-q)}{(r-p)}(x-p) + q - u\right)^2 = w^2$$

so the x-coordinates of the intersection points X and Y are zeros of a quadratic polynomial over K_{i-1}. The same holds for the y-coordinates.

We may now deduce an algebraic consequence of the existence of a construction for a given point.

5.2 Theorem. *If $r = (x, y)$ is constructible from a subset P_0 of $\mathbf{R}^2$, and if K_0 is the subfield of $\mathbf{R}$ generated by the coordinates of the points of P_0, then the degrees*

$$[K_0(x):K_0] \quad and \quad [K_0(y):K_0]$$

are powers of 2.

Proof. We use the notation which we have been accumulating. By 5.1 and 4.3 we have

$$[K_{i-1}(x_i):K_{i-1}] = 1 \text{ or } 2.$$

(The value 2 occurs if the quadratic polynomial over K_{i-1} of which x_i is a zero is irreducible; otherwise the value is 1.) Similarly

$$[K_{i-1}(y_i):K_{i-1}] = 1 \text{ or } 2.$$

Therefore

$$[K_{i-1}(x_i, y_i):K_{i-1}]$$
$$= [K_{i-1}(x_i, y_i):K_{i-1}(x_i)][K_{i-1}(x_i):K_{i-1}]$$
$$= 1, 2, \text{ or } 4.$$

(The value 4 never arises, as the reader may care to prove. This observation is not required for our argument.)

Hence $[K_i:K_{i-1}]$ is a power of 2. By induction (cf. Exercise 4.5) we see that $[K_n:K_0]$ is a power of 2. But since

$$[K_n:K_0(x)][K_0(x):K_0] = [K_n:K_0]$$

it follows that $[K_0(x):K_0]$ is a power of 2. Similarly $[K_0(y):K_0]$ is a power of 2.

Impossibility proofs

We shall now apply the above theory to prove that there do not exist ruler-and-compass constructions for the three classical problems mentioned in the introduction to this chapter. (For the technical drawing expert we emphasize that we are discussing *exact* constructions. There are many *approximate* constructions for trisecting the angle, for instance; but no exact methods.)

5.3 Theorem. *The cube cannot be duplicated using ruler-and-compass constructions.*

Proof. We are given a cube, and hence a side of the cube, which we may take to be the unit interval on the x-axis. Therefore we may assume that $P_0 = \{(0, 0), (1, 0)\}$ so that $K_0 = \mathbf{Q}$. If we could duplicate the cube then we could construct the point $(\alpha, 0)$ where $\alpha^3 = 2$. Therefore by 5.2 $[\mathbf{Q}(\alpha):\mathbf{Q}]$ would be a power of 2. But α is a zero of the polynomial $t^3 - 2$ over $\mathbf{Q}$, and this is irreducible over $\mathbf{Q}$ by Eisenstein's criterion. Hence $t^3 - 2$ is the minimum polynomial of α over $\mathbf{Q}$, and by 4.3 $[\mathbf{Q}(\alpha):\mathbf{Q}] = 3$. This is a contradiction. Therefore the cube cannot be duplicated.

5.4 Theorem. *The angle $\pi/3$ cannot be trisected using ruler-and-compass constructions.*

Proof. To construct an angle trisecting $\pi/3$ is equivalent to constructing the point $(\alpha, 0)$ given $(0, 0)$ and $(1, 0)$, where $\alpha = \cos(\pi/9)$. (See Fig. 4.)

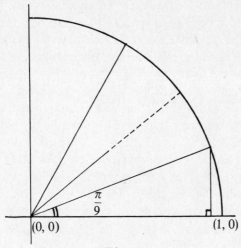

Fig. 4

From this we could construct $(\beta, 0)$ where $\beta = 2\cos(\pi/9)$. From elementary trigonometry we recall the formula

$$\cos(3\theta) = 4\cos^3(\theta) - 3\cos(\theta).$$

If we put $\theta = \pi/9$ then $\cos(3\theta) = \frac{1}{2}$, and we find that

$$\beta^3 - 3\beta - 1 = 0.$$

Now $f(t) = t^3 - 3t - 1$ is irreducible over $\mathbf{Q}$, since $f(t+1) = t^3 + 3t^2 - 3$ is irreducible by Eisenstein's criterion. As in the previous theorem we have $[\mathbf{Q}(\beta):\mathbf{Q}] = 3$, a contradiction.

5.5 Theorem. *The circle cannot be squared using ruler-and-compass constructions.*

Proof. Such a construction is equivalent to one of the point $(0, \sqrt{\pi})$ from $\{(0,0)\,(1,0)\}$. From this we can easily construct

$(0, \pi)$. So $[\mathbf{Q}(\pi):\mathbf{Q}]$ is a power of 2, and in particular π is *algebraic over* $\mathbf{Q}$. On the other hand there is a famous theorem of Lindemann which asserts that π is *not* algebraic over $\mathbf{Q}$. The theorem follows.

We shall prove Lindemann's theorem in the next chapter. The proof involves ideas off the main track of the book, and has therefore been segregated into a separate chapter. The reader who is willing to take it on trust can skip the proof if he wishes; the results are not used anywhere else in the book.

Instead of the traditional ruler and compasses, we may consider other possible methods of construction. Thus in 1672 Mohr showed that every construction performable with ruler and compasses can be performed with compasses alone (provided we agree that a line is constructed when two points on it are given); this result is more commonly attributed to Mascheroni (1797). In 1818 Brianchon considered constructions with ruler only. Poncelet suggested that instead of compasses a single fixed circle, *together with its centre*, should suffice; and this idea was taken up by Steiner in 1833. If the centre of the circle is not known, then fewer constructions are possible. Hilbert asked how many circles must be given in order for the centre of one of them to be constructed using ruler alone. Cauer, in 1912, showed that for two general circles this is impossible, but can be done if they cut, touch, or are concentric. At about the same time Grossmann showed that three linearly independent circles suffice for geometric constructions with only a ruler. It is also known that all ruler-and-compass constructions can be performed using a two-edged ruler, whose edges are either parallel or meet at a point. (References for all these results may be found in Klein [42].)

With additional instruments, more constructions are possible. Thus with a marked ruler the angle can be trisected (see Exercise 5.3). A device for dividing the angle into arbitrarily many equal parts can be found in Cundy and Rollett [9].

Exercises

5.1 Express in the language of this chapter methods of constructing, by ruler and compasses:
(a) The perpendicular bisector of a line.
(b) The points trisecting a line.
(c) Division of a line into n equal parts.
(d) The tangent to a circle at a given point.
(e) Common tangents to two circles.

5.2 Estimate the degrees of the corresponding field extensions by giving reasonably good upper bounds.

5.3 Verify the following construction for trisecting an angle using a marked ruler (Fig. 5). The ruler has marked on it two points distance r apart.

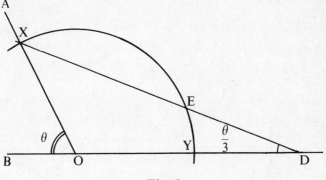

Fig. 5

Given $\angle AOB = \theta$ draw a circle centre O with radius r, cutting OA at X, OB at Y. Place the ruler with its edge through X and one mark on the line OY at D; slide it until the other marked point lies on the circle at E. Then $\angle EDO = \theta/3$.

5.4 Can the angle $2\pi/5$ be trisected using ruler and compasses?

5.5 Show that it is impossible to construct a regular 9-gon using ruler and compasses.

5.6 By considering a formula for $\cos(5\theta)$ find a construction for the regular pentagon.

5.7 Prove that the angle θ can be trisected by ruler and compasses if and only if the polynomial

$$4t^3 - 3t - \cos(\theta)$$

is reducible over $\mathbf{Q}(\cos(\theta))$.

5.8 Discuss the quinquisection (division into 5 equal parts) of angles.

5.9 Verify the following approximate construction for π due to Ramanujan [24] p. 35 (Fig. 6).

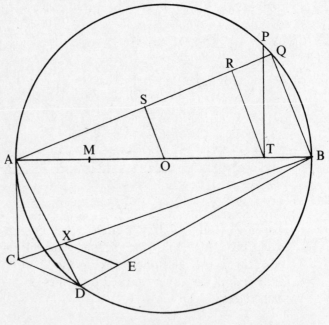

Fig. 6

Let AB be the diameter of a circle centre O. Bisect AO at M, trisect OB at T. Draw TP perpendicular to AB meeting the circle at P. Draw BQ = PT, and join AQ.

Draw OS, TR parallel to BQ. Draw AD = AS, and AC = RS tangential to the circle at A. Join BC, BD, CD. Make BE = BM. Draw EX parallel to CD. Then the square on BX has approximately the same area as the circle.

(You will need to know that π is approximately 355/113. This approximation is first found in the works of the Chinese astronomer Tsu Ch'ung Ching in about 450 A.D.)

5.10 Mark the following true or false.

(a) There exist constructions trisecting the angle to an arbitrary degree of approximation.

(b) Such constructions are sufficient for practical purposes but insufficient for mathematical ones.

(c) The coordinates of a constructible point lie in a subfield of **R** whose degree over the subfield generated by the coordinates of the given points is a power of 2.

(d) The angle π cannot be trisected using ruler and compasses.

(e) A line of length π cannot be constructed from $\{(0, 0), (1, 0)\}$ using ruler and compasses.

(f) It is impossible to triplicate the cube by ruler and compasses.

(g) The real number π is transcendental over **Q**.

(h) The real number π is transcendental over **R**.

(i) If $(0, \alpha)$ cannot be constructed from $\{(0, 0), (1, 0)\}$ by ruler and compasses then α is transcendental over **Q**.

(j) Geometrical problems cannot always be solved geometrically.

CHAPTER SIX

Transcendental numbers

The material in this chapter is not used elsewhere and may be omitted if so desired.

To complete the proof of the impossibility of squaring the circle (and so crown 3000 years of mathematical effort) we must prove that π is transcendental over $\mathbf{Q}$. (In this chapter the word 'transcendental' will be understood to mean transcendental over $\mathbf{Q}$.) The proof we shall give is analytical in character, which should not be surprising since π is best defined analytically. The techniques involve integration, differentiation, and some manipulation of inequalities, together with a healthy lack of respect for apparently complicated expressions.

It is not at all obvious that transcendental numbers exist within the complex field. That they do was first proved by Liouville in 1844 by considering the approximation of reals by rationals. It transpires that algebraic numbers cannot be approximated by rationals with more than a certain 'speed' (see Exercises 6.6–6.8). To find a transcendental number reduces to finding a number which can be approximated more rapidly than the known bound for algebraic numbers. Liouville showed that this was the case for the real number

$$\xi = \sum_{n=1}^{\infty} 10^{-n!}$$
$$= 0 \cdot 110001000000000000000000000100\ldots$$

However, no 'naturally occurring' number was proved transcendental until Hermite, in 1873, proved that e, the 'base of natural logarithms', was. Using similar methods Lindemann demonstrated the transcendence of π in 1882.

Meanwhile Cantor, in 1874, had produced a revolutionary proof of the existence of transcendental numbers, *without actually constructing any.* His proof used set-theoretic methods, and was one of the earliest triumphs of Cantor's theory of infinite cardinals. The mathematical world viewed it with great suspicion. (See Exercises 6.1–6.4).

We shall prove four theorems in this chapter. In each case the proof procedes by contradiction, and the final blow is dealt by the following simple result:

6.1 Lemma. *A function $f : \mathbf{Z} \to \mathbf{Z}$ which tends to zero as the variable tends to infinity must be eventually zero.*

Proof. Since $f(n) \to 0$ as $n \to \infty$ we have

$$|f(n) - 0| < \tfrac{1}{2}$$

if $n \geq N$, for some integer N. Since $f(n)$ is an integer this implies that $f(n) = 0$ for $n \geq N$.

Irrationality

Lindemann's proof is ingenious and intricate. To prepare the way we shall first prove some simpler theorems of the same general type. The *results* of these are not needed for Lindemann's proof; but familiarity with the *ideas* is. Our first theorem was initially proved by Lambert in 1770 using continued fractions, although it is often credited to Legendre.

6.2 Theorem. *The real number π is irrational.*

Proof. Consider the integral

$$I_n = \int_{-1}^{+1} (1 - x^2)^n \cos(\alpha x)\, dx.$$

Integrating by parts we find that

$$\alpha^2 I_n = 2n(2n-1)I_{n-1} - 4n(n-1)I_{n-2} \qquad (1)$$

if $n \geq 2$. By induction on n it follows that

$$\alpha^{2n+1} I_n = n!(P \sin(\alpha) + Q \cos(\alpha)) \qquad (2)$$

where P and Q are polynomials in α of degree $< 2n+1$ with integer coefficients. The term $n!$ comes from the factor $2n(2n-1)$ of equation (1).

Assume now, for a contradiction, that π is rational, so that $\pi = a/b$ where $a, b \in \mathbf{Z}$ and $b \neq 0$. Let $\alpha = \pi/2$ in (2). Then

$$J_n = b^{2n+1} I_n/n!$$

is an integer. Now

$$J_n = \frac{b^{2n+1}}{n!} \int_{-1}^{+1} (1-x^2)^n \cos \frac{\pi}{2}x \ dx.$$

The integrand is > 0 for $-1 < x < 1$ so that $J_n > 0$. Hence $J_n \neq 0$ for all n. But

$$|J_n| \leq \frac{|b|^{2n+1}}{n!} \int_{-1}^{+1} \cos \frac{\pi}{2}x \ dx$$

$$\leq C|b|^{2n+1}/n!$$

where C is a constant. Hence $J_n \to 0$ as $n \to \infty$. This contradicts Lemma 6.1, so our assumption that π is rational is false.

Our next result was proved by Legendre in 1794 in his *Éléments de Géométrie* (which as we have remarked in the biographical sketch greatly influenced the young Galois).

6.3 Theorem. *The real number π^2 is irrational.*

Proof. Assume if possible that $\pi^2 = a/b$ where $a, b \in \mathbf{Z}$ and $b \neq 0$. Define

$$f(x) = x^n(1-x)^n/n!$$

and

$$G(x) = b^n\{\pi^{2n}f(x) - \pi^{2n-2}f''(x) + \cdots + (-1)^n\pi^0 f^{(2n)}(x)\}$$

where the superscripts indicate derivatives. We claim that any derivative of f takes integer values at 0 and 1. Recall Leibniz's rule for differentiating a product:

$$\frac{d^m}{dx^m}(uv) = \sum \binom{m}{r}\frac{d^r u}{dx^r}\frac{d^{m-r}v}{dx^{m-r}}.$$

If both factors x^n or $(1-x)^n$ are differentiated fewer than n times the value of the corresponding term is 0 whenever $x = 0$ or 1. If one factor is differentiated n or more times then the denominator $n!$ is cancelled out. Hence $G(0)$ and $G(1)$ are integers. Now

$$\frac{d}{dx}\{G'(x)\sin(\pi x) - \pi G(x)\cos(\pi x)\}$$

$$= \{G''(x) + \pi^2 G(x)\}\sin(\pi x)$$

$$= b^n \pi^{2n+2} f(x)\sin(\pi x)$$

since $f(x)$ is a polynomial in x of degree $2n$, so that $f^{(2n+2)}(x) = 0$. And this expression is equal to

$$\pi^2 a^n \sin(\pi x) f(x).$$

Therefore

$$\pi \int_0^1 a^n \sin(\pi x) f(x)\,dx = \left[\frac{G'(x)\sin(\pi x)}{\pi} - G(x)\cos(\pi x)\right]_0^1$$

$$= G(0) + G(1)$$

which is an integer. As before the integral is not zero. But

$$\left|\int_0^1 a^n \sin(\pi x) f(x)\,dx\right| \le |a|^n \int_0^1 |\sin(\pi x)||f(x)|\,dx$$

$$\le |a|^n \int_0^1 \frac{|x^n(1-x)^n|}{n!}\,dx$$

$$\le \frac{1}{n!}\int_0^1 |(ax)^n(1-x)^n|\,dx$$

which tends to 0 as n tends to ∞. The usual contradiction completes the proof.

Transcendence of e

We move on from irrationality to the far more elusive transcendence. Hermite's original proof was simplified by Weierstrass, Hilbert, Hurwitz, and Gordan; and it is the simplified proof which we give here. (Similarly for the proof of Lindemann's theorem.)

6.4 Theorem (*Hermite*). *The real number* e *is transcendental.*

Proof. Assume that e is not transcendental. Then

$$a_m e^m + \cdots + a_i e + a_0 = 0$$

where without loss of generality we may suppose that $a_i \in \mathbf{Z}$ for all i and $a_0 \neq 0$. Define

$$f(x) = \frac{x^{p-1}(x-1)^p(x-2)^p \cdots (x-m)^p}{(p-1)!}$$

where p is an arbitrary prime number. Now f is a polynomial in x of degree $mp + p - 1$. Put

$$F(x) = f(x) + f'(x) + \cdots + f^{(mp+p-1)}(x)$$

and note that $f^{(mp+p)}(x) = 0$. We calculate:

$$\frac{\mathrm{d}}{\mathrm{d}x}\{e^{-x}F(x)\} = e^{-x}\{F'(x) - F(x)\}$$

$$= -e^{-x}f(x).$$

Hence for any j

$$a_j \int_0^j e^{-x}f(x)\,\mathrm{d}x = a_j\left[-e^{-x}F(x)\right]_0^j$$

$$= a_j F(0) - a_j e^{-j}F(j).$$

Multiplying by e^j and summing over $j = 0, 1, \cdots, m$ we get

$$\sum_{j=0}^{m} \left(a_j e^j \int_0^j e^{-x} f(x) \, dx \right) = F(0) \sum_{j=0}^{m} a_j e^j - \sum_{j=0}^{m} a_j F(j)$$

$$= -\sum_{j=0}^{m} \sum_{i=0}^{mp+p-1} a_j f^{(i)}(j) \qquad (1)$$

from the equation supposedly satisfied by e.

We claim now that each $f^{(i)}(j)$ is an integer, and that this integer is divisible by p unless $j = 0$ and $i = p-1$. We use Leibniz's rule again; the only non-zero terms arising when $j \neq 0$ come from the factor $(x-j)^p$ being differentiated exactly p times. Since $p!/(p-1)! = p$, all such terms are integers divisible by p. In the exceptional case $j = 0$ the only non-zero term is when $i = p-1$ and then

$$f^{(p-1)}(0) = (-1)^p \cdots (-m)^p.$$

The value of (1) is therefore

$$Kp + a_0(-1)^p \cdots (-m)^p$$

for some $K \in \mathbf{Z}$. Now if $p > \max(m, |a_0|)$ then the integer $a_0(-1)^p \cdots (-m)^p$ is not divisible by p. So for sufficiently large primes p the value of (1) is an integer not divisible by p, hence *not zero*.

Now we estimate the integral. If $0 \leq x \leq m$ then

$$|f(x)| \leq m^{mp+p-1}/(p-1)!$$

so that

$$\left| \sum_{j=0}^{m} a_j e^j \int_0^j e^{-x} f(x) \, dx \right| \leq \sum_{j=0}^{m} |a_j e^j| \int_0^j \frac{m^{mp+p-1}}{(p-1)!} \, dx$$

$$\leq \sum_{j=0}^{m} |a_j e^j| j \cdot \frac{m^{mp+p-1}}{(p-1)!}$$

which tends to 0 as p tends to ∞.

This is the usual contradiction. Therefore e is transcendental.

Transcendence of π

Our proof that π is transcendental involves the same sort of trickery as the previous results, but of a far more elaborate nature. At several points in the proof we shall use the properties of symmetric polynomials (Chapter 2).

6.5 Theorem. (*Lindemann*). *The real number π is transcendental.*

Proof. Suppose for a contradiction that π is a zero of some non-zero polynomial over $\mathbf{Q}$. Then so is $i\pi$, where $i = \sqrt{-1}$. Let $\theta_1(x) \in \mathbf{Q}[x]$ be a polynomial with zeros $\alpha_1 = i\pi$, $\alpha_2, \cdots, \alpha_n$. By a famous theorem of Euler we have

$$e^{i\pi} + 1 = 0$$

so that

$$(e^{\alpha_1} + 1)(e^{\alpha_2} + 1) \cdots (e^{\alpha_n} + 1) = 0. \tag{2}$$

We now construct a polynomial with integer coefficients whose zeros are the exponents $\alpha_{i_1} + \cdots + \alpha_{i_r}$ of e which appear in the expansion of the product (2). For example, terms of the form

$$e^{\alpha_s} . e^{\alpha_t} . 1 . 1 . 1 \ldots . 1$$

give rise to exponents $\alpha_s + \alpha_t$. Taken over all pairs s, t we get $\alpha_1 + \alpha_2, \cdots, \alpha_{n-1} + \alpha_n$. The elementary symmetric polynomials of these are symmetric in $\alpha_1, \cdots, \alpha_n$, so by 2.9 they can be expressed as polynomials in the elementary symmetric polynomials of $\alpha_1, \cdots, \alpha_n$. These in turn are expressible in terms of the coefficients of the polynomial θ_1 whose zeros are $\alpha_1, \cdots, \alpha_n$. Hence the pairs $\alpha_s + \alpha_t$ satisfy a polynomial equation $\theta_2(x) = 0$ where θ_2 has rational coefficients. Similarly the sums of k α's are zeros of a polynomial $\theta_k(x)$ over $\mathbf{Q}$. Then

$$\theta_1(x)\theta_2(x) \cdots \theta_n(x)$$

is a polynomial over $\mathbf{Q}$ whose zeros are the exponents of e in the expansion of (2). Dividing by a suitable power of x

and multiplying by a suitable integer we obtain a polynomial $\theta(x)$ over **Z**, whose zeros are the *non-zero* exponents $\beta_1, \cdots, \beta_r$ of e in the expansion of (2). Now (2) takes the form

$$e^{\beta_1} + \cdots + e^{\beta_r} + e^0 + \cdots + e^0 = 0,$$

that is,

$$e^{\beta_1} + \cdots + e^{\beta_r} + k = 0 \qquad (3)$$

where $k \in$ **Z**. The term $1.1\ldots\ldots1$ occurs in the expansion, so that $k > 0$.

Suppose that

$$\theta(x) = cx^r + c_1 x^{r-1} + \cdots + c_r.$$

Now $c_r \neq 0$ since 0 is not a zero of θ. Define

$$f(x) = \frac{c^s x^{p-1} \{\theta(x)\}^p}{(p-1)!}$$

where $s = r(p-1)$ and p is any prime number. Define also

$$F(x) = f(x) + f'(x) + \cdots + f^{(s+p+r-1)}(x)$$

and note that $f^{(s+p+r)}(x) = 0$. As before we have

$$\frac{\mathrm{d}}{\mathrm{d}x}\{e^{-x}F(x)\} = -e^{-x}f(x).$$

Hence

$$e^{-x}F(x) - F(0) = -\int_0^x e^{-y}f(y)\,\mathrm{d}y.$$

Putting $y = \lambda x$ we get

$$F(x) - e^x F(0) = -x\int_0^1 e^{(1-\lambda)x}f(\lambda x)\,\mathrm{d}\lambda.$$

Let x range over $\beta_1, \cdots, \beta_r$ and sum: from (3) we obtain

$$\sum_{j=1}^r F(\beta_j) + kF(0) = -\sum_{j=1}^r \beta_j \int_0^1 e^{(1-\lambda)\beta_j}f(\lambda\beta_j)\,\mathrm{d}\lambda. \qquad (4)$$

We claim that for all sufficiently large p the left-hand side of

(4) is a non-zero integer. Now

$$\sum_{j=1}^{r} f^{(t)}(\beta_j) = 0$$

if $0 < t < p$. Each derivative $f^{(t)}(\beta_j)$ with $t \geq p$ has a factor p since we must differentiate $\{\theta(x)\}^p$ at least p times to obtain a non-zero term. For any such t,

$$\sum_{j=1}^{r} f^{(t)}(\beta_j)$$

is a symmetric polynomial in the β_j of degree $\leq s$. Thus it is a polynomial of degree $\leq s$ in the coefficients c_i/c, by 2.9. The factor c^s in the definition of $f(x)$ makes this into an integer. So for $t \geq p$

$$\sum_{j=1}^{r} f^{(t)}(\beta_j) = pk_t$$

for suitable $k_t \in \mathbf{Z}$. Now we look at $F(0)$. We have

$$f^{(t)}(0) = \begin{cases} 0 & (t \leq p-2) \\ c^s c_r^p & (t = p-1) \\ l_t p & (t \geq p) \end{cases}$$

for suitable $l_t \in \mathbf{Z}$. Consequently the left-hand side of (4) is

$$Kp + kc^s c_r^p$$

for some $K \in \mathbf{Z}$. Now $k \neq 0$, $c \neq 0$, and $c_r \neq 0$. If we take

$$p > \max(k, |c|, |c_r|)$$

then the left-hand side of (4) is an integer not divisible by p, so is non-zero.

The last part of the proof is routine: we estimate the size of the right-hand side of (4). Now

$$|f(\lambda\beta_j)| \leq \frac{|c|^s |\beta_j|^{p-1} (m(j))^p}{(p-1)!}$$

where

$$m(j) = \sup_{0 \leq \lambda \leq 1} |\theta(\lambda\beta_j)|.$$

Therefore

$$\left| -\sum_{j=1}^{r} \beta_j \int_0^1 e^{(1-\lambda)\beta_j} f(\lambda \beta_j)\, \mathrm{d}\lambda \right|$$

$$\leq \sum_{j=1}^{r} \frac{|\beta_j|^p |c^s| |m(j)|^p B}{(p-)!}$$

where

$$B = \left| \max_j \int_0^1 e^{(1-\lambda)\beta_j}\, \mathrm{d}\lambda \right|.$$

Thus the expression tends to 0 as p tends to ∞. By the standard contradiction, π is transcendental.

Exercises

The first 4 exercises outline Cantor's proof of the existence of transcendental numbers.

6.1 Prove that **R** is uncountable.

6.2 Define the *height* of a polynomial

$$f(t) = a_0 + \cdots + a_n t^n \in \mathbf{Z}[t]$$

to be

$$h(f) = n + |a_0| + \cdots + |a_n|.$$

Prove that there are only a finite number of polynomials over **Z** of given height h.

6.3 Show that any algebraic number satisfies a polynomial equation over **Z**. Using 6.2 show that the algebraic numbers form a countable set.

6.4 Combine 6.1 and 6.3 to show that transcendental numbers exist.

6.5 Thomas Hobbes (a philosopher) suggested the following construction for squaring the circle (Fig. 7).

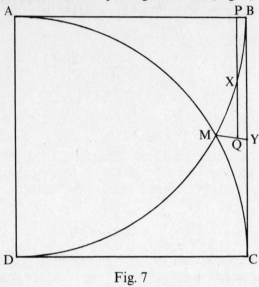

Fig. 7

Inside a unit square ABCD draw two quadrants of
circles, meeting at M. Bisect arc BM at X. Draw PQ
through X parallel to BC, so that PX = XQ. Join
MQ and produce to meet BC at Y. Then BY is
exactly equal to the length of arc BM.
  Find the mistake.

The next three exercises give Liouville's proof of the
existence of transcendental numbers.

6.6 Suppose that x is irrational and that

$$f(x) = a_n x^n + \cdots + a_0 = 0$$

where $a_0, \cdots, a_n \in \mathbf{Z}$. Show that if $p, q \in \mathbf{Z}$ and $q \neq 0$,
and $f(p/q) \neq 0$, then

$$|f(p/q)| \geq 1/q^n.$$

6.7 Now suppose that $x - 1 < p/q < x + 1$ and p/q is nearer to x than any other zero of f. There exists M such that $|f'(y)| < M$ if $x-1 < y < x+1$. Use the mean value theorem to show that

$$|p/q - x| \geq M^{-1} q^{-n}.$$

Hence show that for any $r > n$ and $K > 0$ there exist only finitely many p and q such that

$$|p/q - x| < Kq^{-r}.$$

6.8 Use this result to prove that $\sum\limits_{n=1}^{\infty} 10^{-n!}$ is transcendental.

6.9 Mark the following true or false.
(a) π is irrational
(b) All irrational numbers are transcendental.
(c) Any rational multiple of π is transcendental.
(d) Sometimes the only way to prove a theorem is to pull rabbits out of hats.
(e) e is irrational.
(f) If α and β are transcendental then so is $\alpha + \beta$.
(g) Transcendental numbers form a subring of $\mathbf{C}$.
(h) The field $\mathbf{Q}(\pi)$ is isomorphic to $\mathbf{Q}(t)$ for any indeterminate t.
(i) $\mathbf{Q}(\pi)$ and $\mathbf{Q}(e)$ are non-isomorphic fields.
(j) $\mathbf{Q}(\pi)$ is isomorphic to $\mathbf{Q}(\pi^2)$.

CHAPTER SEVEN

The idea behind Galois Theory

It will take some time before we can prove the main theorem of Galois Theory. To avoid losing the line of argument among a mass of technicalities we shall here outline the theorem we wish to prove and the steps required to prove it.

We have already associated a vector space with each field extension. For some problems this is too coarse an instrument; it measures the *size* but not the *shape* (so to speak). Galois went deeper into the structure. With any polynomial he associated a group of permutations of its zeros, now called the *Galois group* in his honour. (In the modern field-theoretic treatment this group is defined in terms of the corresponding field extension; but Galois had the basic idea.) What makes his work so astonishing is that at the time the group concept existed only in rudimentary form. Galois was the first to recognize its importance.

How can we find a suitable group? Suppose we have an extension $L:K$. Then the set of all automorphisms of L is certainly a group under composition of maps, but not a very interesting one. It tells us nothing about the status of K, for all subfields of L are on the same footing relative to the automorphisms of L. We bring K into the picture by using a group related both to L and to K.

Definition. Let K be a subfield of the field L. An automorphism α of L is a K-*automorphism* of L if

$$\alpha(k) = k$$

for all $k \in K$.

Effectively this makes α an automorphism of the *extension* rather than just the large field. The idea of considering automorphisms of a mathematical object relative to a sub-object is a useful general method; it falls within the scope of the famous *Erlanger Programm* of Felix Klein. Klein's idea was to consider every 'geometry' as the theory of invariants of a particular transformation group. Thus Euclidean geometry is the study of invariants of the group of distance-preserving transformations of the plane; projective geometry arises if we allow projective transformations; topology comes from the group of all continuous maps possessing continuous inverses (called *homeomorphisms* or *topological transformations*). According to this interpretation any field extension is a geometry, and we are simply studying the geometrical figures.

The pivot upon which the whole theory turns is a result which is not in itself hard to prove. It is a 'maxim tremendous but trite'‡ whose *implications* far exceed its *content*.

7.1 Theorem. *If $L:K$ is a field extension then the set of all K-automorphisms of L forms a group under composition of maps.*

Proof. Suppose that α and β are K-automorphisms of L. Then $\alpha\beta$ is clearly an automorphism; further if $k \in K$ then $\alpha\beta(k) = \alpha(k) = k$, so that $\alpha\beta$ is a K-automorphism. The identity map on L is obviously a K-automorphism. Finally, α^{-1} is an automorphism of L, and for any $k \in K$ we have $k = \alpha^{-1}\alpha(k) = \alpha^{-1}(k)$ so that α^{-1} is a K-automorphism. Composition of maps is associative, so the set of all K-automorphisms of L is a group.

‡ Quoted from *The Hunting of the Snark* by Lewis Carroll.

Definition. The *Galois group* $\Gamma(L:K)$ of the extension $L:K$ is the group of all K-automorphisms of L under composition of maps.

Examples

1 The extension $\mathbf{C}:\mathbf{R}$. Suppose that α is an $\mathbf{R}$-automorphism of $\mathbf{C}$. Let $j = \alpha(i)$ where $i = \sqrt{-1}$. Then

$$j^2 = (\alpha(i))^2 = \alpha(i^2) = \alpha(-1) = -1$$

since $\alpha(r) = r$ for all $r \in \mathbf{R}$. Hence either $j = i$ or $j = -i$. Now for any $x, y \in \mathbf{R}$ we have

$$\alpha(x+iy) = \alpha(x)+\alpha(i)\alpha(y)$$
$$= x+jy.$$

Thus we have two candidates for $\mathbf{R}$-automorphisms:

$$\alpha_1 : x+iy \rightarrow x+iy$$
$$\alpha_2 : x+iy \rightarrow x-iy.$$

Now α_1 is the identity, and thus is an $\mathbf{R}$-automorphism of $\mathbf{C}$. The map α_2 is known as *complex conjugation* and can be shown to be an $\mathbf{R}$-automorphism as follows:

$$\alpha_2((x+iy)+(u+iv)) = (x+y)-(u+v)i$$
$$= \alpha_2(x+iy)+\alpha_2(u+iv).$$
$$\alpha_2((x+iy)(u+iv)) = \alpha_2(xu-yv+i(xv+yu))$$
$$= xu-yv-i(xv+yu)$$
$$= (x-iy)(u-iv)$$
$$= \alpha_2(x+iy)\alpha_2(u+iv).$$

Thus α_2 is an automorphism. And

$$\alpha_2(x+0i) = x-0i = x$$

so that α_2 is an $\mathbf{R}$-automorphism.

Obviously $\alpha_2^2 = \alpha_1$, so that the Galois group $\Gamma(\mathbf{C}:\mathbf{R})$ is a cyclic group of order 2.

2 Let c be the real cube root of 2, and consider $\mathbf{Q}(c):\mathbf{Q}$. If α is a $\mathbf{Q}$-automorphism of $\mathbf{Q}(c)$ then

$$(\alpha(c))^3 = \alpha(c^3) = \alpha(2) = 2.$$

Since $\mathbf{Q}(c) \subseteq \mathbf{R}$ we must have $\alpha(c) = c$. Hence α is the identity map and $\Gamma(\mathbf{Q}(c):\mathbf{Q})$ has order 1.

Although it is easy to prove that the set of all K-automorphisms of a field L forms a group, that fact alone does not significantly advance the subject. To be of any use, the Galois group must reflect aspects of the structure of $L:K$. Galois made the discovery (expressed by him in terms of polynomials, of course) that *under certain extra hypotheses* there is a 1–1 correspondence between:
(1) Subgroups of the Galois group of $L:K$,
(2) Subfields M of L such that $K \subseteq M$.
It is also true that this correspondence reverses inclusion relations. We shall return to this point in a moment, but first we shall explain how the correspondence goes.

If $L:K$ is a field extension we shall call a field M such that $K \subseteq M \subseteq L$ an *intermediate* field. To each intermediate field M we associate the group $M^* = \Gamma(L:M)$ of all M-automorphisms of L. Thus K^* is the whole Galois group, $L^* = 1$ (the identity map on L). Clearly if $M \subseteq N$ then $M^* \supseteq N^*$, for any map fixing the elements of N certainly fixes the elements of M.

Conversely, to each subgroup H of $\Gamma(L:K)$ we associate the set $H^\dagger$ of all elements $x \in L$ such that $\alpha(x) = x$ for all $\alpha \in H$. That this *is* an intermediate field follows from:

7.2 Lemma. *If H is a subgroup of $\Gamma(L:K)$ then $H^\dagger$ is a subfield of L containing K.*

Proof. Let $x, y \in H^\dagger$, and $\alpha \in H$. Then $\alpha(x+y) = \alpha(x)+\alpha(y) = x+y$. Similarly $H^\dagger$ is closed under the other field operations, so is a subfield of L. Since $\alpha \in \Gamma(L:K)$ we have $\alpha(k) = k$ for all $k \in K$, so that $K \subseteq H^\dagger$.

Definition. With the above notation, $H^\dagger$ is the *fixed field* of H.

Clearly if $H \subseteq G$ then $H^\dagger \supseteq G^\dagger$. It is also easy to verify that if M is an intermediate field and H is a subgroup of the Galois group then

$$M \subseteq M^{*\dagger}$$
$$H \subseteq H^{\dagger*} \tag{1}$$

since every element of M is left fixed by every automorphism fixing all of M; and every element of H fixes those elements left fixed by all of H. Examples such as 2 above show that these inclusions are not always equalities, for in that example we have

$$\mathbf{Q}^{*\dagger} = \mathbf{Q}(c).$$

If we let $\mathscr{F}$ denote the set of intermediate fields, and $\mathscr{G}$ the set of subgroups of the Galois group, then the situation is that we have two maps

$$*: \mathscr{F} \to \mathscr{G}$$
$$\dagger: \mathscr{G} \to \mathscr{F}$$

which reverse inclusions and satisfy (1). Galois's ideas can be interpreted as giving *conditions under which* * *and* $\dagger$ *are mutual inverses*, setting up a bijection between $\mathscr{F}$ and $\mathscr{G}$. The extra conditions needed are called *separability* and *normality*. We shall discuss them in the next chapter.

Exercises

7.1 Find the K-automorphisms of L for the following extensions $L:K$.
(a) $\mathbf{Q}(\sqrt{2}):\mathbf{Q}$
(b) $\mathbf{Q}(\alpha):\mathbf{Q}$ where α is the real 5th root of 7
(c) $\mathbf{Q}(\sqrt{2}, \sqrt{3}):\mathbf{Q}$.

7.2 Compute the corresponding Galois groups for these extensions.

7.3 In which of the above cases is the Galois correspondence between $\mathscr{F}$ and $\mathscr{G}$ a bijection?

7.4 Let K be the field of four elements defined in Exercise 1.6, and let P be its prime subfield. What is the Galois group of $K:P$? Is the Galois correspondence a bijection?

7.5 Let $K(\alpha):K$ be a simple algebraic extension, and suppose that γ is an element of the Galois group. Show that $\gamma(\alpha)$ has the same minimum polynomial as α over K. Hence show that the Galois group acts as permutations of the zeros of this polynomial.

7.6 Find all intermediate fields for the extension $\mathbf{Q}(\sqrt{2}, \sqrt{3}, \sqrt{5}):\mathbf{Q}$. Find the Galois group. Compare.

7.7 Mark the following true or false.
  (a) Every K-automorphism of L is an automorphism of L.
  (b) Every L-automorphism of L is the identity.
  (c) The Galois group of $L:K$ is a cyclic group.
  (d) The Galois group of $\mathbf{C}:\mathbf{R}$ is abelian.
  (e) The maps * and † are always mutual inverses.
  (f) The maps * and † preserve inclusions.
  (g) If $\Gamma(L:K) = 1$ then $L = K$.
  (h) If $L = K$ then $\Gamma(L:K) = 1$.
  (i) $K(t)$ has only one K-automorphism.
  (j) Galois groups are easier to define than to compute.

CHAPTER EIGHT

Normality and separability

In this chapter we shall define the important concepts of *normality* and *separability* and develop some key results concerning them.

We shall also take a close look at the phenomenon whereby a polynomial may have no zeros over one field, but has zeros over a larger field. For example $t^2 + 1$ has no zeros over **R**, but has zeros $\pm i$ over **C**. It will turn out that every polynomial can be resolved into a product of linear factors (and hence has its full complement of zeros) if the ground field is extended to a suitable *splitting field*. We shall study these splitting fields, and show that they are closely related to the idea of normality.

In order to give examples we shall make use of a special property of the field **C** of complex numbers; a result often known as the *Fundamental Theorem of Algebra*, namely: *Every polynomial over **C** can be expressed as a product of linear factors*. We shall prove this result in Chapter 18.

Splitting fields

We begin by defining our terms.

Definition. If K is a field and f is a polynomial over K then f *splits* over K if it can be expressed as a product of linear factors

$$f(t) = k(t - \alpha_1) \cdots (t - \alpha_n)$$

where $k, \alpha_1, \cdots, \alpha_n \in K$.

If this is the case then the zeros of f in K are precisely $\alpha_1, \cdots, \alpha_n$.

If f is a polynomial over K and L is an extension field of K then f is also a polynomial over L, and it makes sense to talk of f splitting *over* L. We show that given K and f we can always construct an extension Σ of K such that f splits over Σ. It is convenient to require in addition that f does not split over any smaller field, so that Σ is as economical as possible.

Definition. The field Σ is a *splitting field* for the polynomial f over the field K if $K \subseteq \Sigma$ and
(1) f splits over Σ,
(2) If $K \subseteq \Sigma' \subseteq \Sigma$ and f splits over Σ' then $\Sigma' = \Sigma$.

The second condition is clearly equivalent to:
(2') $\Sigma = K(\sigma_1, \cdots, \sigma_n)$ where $\sigma_1, \cdots, \sigma_n$ are the zeros of f in Σ.

We construct a splitting field by adjoining to K elements which are to be thought of as the zeros of f. We already know how to do this for irreducible polynomials (Theorem 3.5), so we split f into irreducible factors and work on these separately.

8.1 Theorem. If K is any field and f is any polynomial over K then there exists a splitting field for f over K.

Proof. We use induction on the degree ∂f. If $\partial f = 1$ there is nothing to prove, for f splits over K. If f does not split over K then it has an irreducible factor f_1 of degree > 1. Using 3.5 we adjoin σ_1 to K, where $f(\sigma_1) = 0$. Then in $K(\sigma_1)[t]$ we have $f = (t - \sigma_1)g$ where $\partial g = \partial f - 1$. By induction there is a splitting field Σ for g over $K(\sigma_1)$. But then Σ is a splitting field for f over K.

It would appear at first sight that we might construct different splitting fields for f by varying the choice of irreducible factors. In fact this makes no difference up to isomorphism, and we shall prove that splitting fields (for given f and K) are unique up to isomorphism. This result is not really surprising, but the proof is more complicated than one might hope. The idea is straightforward; use 3.8 and some sort of induction. For technical reasons we use the stronger 3.9. The main point of the proof is embodied in:

8.2 Lemma. *Suppose* $i:K \to K'$ *is an isomorphism of fields. Let f be a polynomial over K and let Σ be any splitting field for f over K. Let L be any extension field of K' such that $i(f)$ splits over L. Then there exists a monomorphism $j:\Sigma \to L$ such that $j|_K = i$.*

Proof. We have the following situation:

$$
\begin{array}{ccc}
K & \to & \Sigma \\
i \downarrow & & \downarrow j \\
K' & \to & L
\end{array}
$$

where j has yet to be found.

We use induction on ∂f. As a polynomial over Σ

$$f(t) = k(t - \sigma_1) \cdots (t - \sigma_n).$$

The minimum polynomial m of σ_1 over K is an irreducible factor of f. Now $i(m)$ divides $i(f)$ which splits over L, so that over L we have

$$i(m) = (t - \alpha_1) \cdots (t - \alpha_r)$$

where $\alpha_1, \cdots, \alpha_r \in L$. Since $i(m)$ is irreducible over K' it must be the minimum polynomial of α_1 over K'. So by 3.9 there is an isomorphism

$$j_1 : K(\sigma_1) \to K'(\alpha_1)$$

such that $j_1|_K = i$ and $j_1(\sigma_1) = \alpha_1$. Now Σ is a splitting field over $K(\sigma_1)$ of the polynomial $g = f/(t - \sigma_1)$. By induction there exists a monomorphism $j:\Sigma \rightarrow L$ such that $j|_{K(\sigma_1)} = j_1$. But then $j|_K = i$ and we are finished.

This enables us to prove the uniqueness theorem.

8.3 Theorem. *Let $i:K \rightarrow K'$ be a field isomorphism. Let T be a splitting field for f over K, T' a splitting field for i(f) over K'. Then there is an isomorphism $j:T \rightarrow T'$ such that $j|_K = i$. In other words the extensions T:K and T':K' are isomorphic.*

Proof. We start with the following diagram.

$$
\begin{array}{ccc}
K & \rightarrow & T \\
i \downarrow & & \downarrow j \\
K' & \rightarrow & T'
\end{array}
$$

We must find j to make the diagram commute. Now by 8.2 there is a *monomorphism* $j:T \rightarrow T'$ such that $j|_K = i$. But $j(T)$ is a splitting field for $i(f)$ over K', and is contained in T'. Since T' is also a splitting field for $i(f)$ over K' we have $j(T) = T'$, so that j is surjective. Hence j is an isomorphism and the theorem follows.

Examples
1 Let $f(t) = (t^2 - 3)(t^3 + 1)$ over $\mathbf{Q}$. We can construct a splitting field for f as follows: inside $\mathbf{C}$ f splits into linear factors

$$f(t) = (t+\sqrt{3})(t-\sqrt{3})(t+1)\left(t - \frac{-1+i\sqrt{3}}{2}\right)\left(t - \frac{-1-i\sqrt{3}}{2}\right)$$

so there exists a splitting field inside $\mathbf{C}$, namely

$$\mathbf{Q}\left(\sqrt{3}, \frac{-1+i\sqrt{3}}{2}\right).$$

This is clearly the same as $\mathbf{Q}(\sqrt{3}, i)$.
2 Let $f(t) = (t^2 - 2t - 2)(t^2 + 1)$ over $\mathbf{Q}$. The zeros of f in $\mathbf{C}$ are $1 \pm \sqrt{3}$, $\pm i$, so a splitting field is afforded by $\mathbf{Q}(1+\sqrt{3}, i)$ which equals $\mathbf{Q}(\sqrt{3}, i)$. This is the same field

as in the previous example, although the two polynomials involved are different.

3 It is even possible to have two distinct *irreducible* polynomials with the same splitting field. For example $t^2 - 3$ and $t^2 - 2t - 2$ both have $\mathbf{Q}(\sqrt{3})$ as a splitting field over $\mathbf{Q}$.

4 Now consider $f(t) = t^2 + t + 1$ over $\mathbf{Z}_2$. This time we cannot use $\mathbf{C}$, so we **must** go back to the basic construction for a splitting field. The ground field $\mathbf{Z}_2$ has two elements, 0 and 1. We note that f is irreducible, so we may adjoin an element ζ such that ζ has minimum polynomial f over $\mathbf{Z}_2$. Then $\zeta^2 + \zeta + 1 = 0$ so that $\zeta^2 = 1 + \zeta$ (characteristic 2!) and the following elements form a field:

$$0, 1, \zeta, 1 + \zeta.$$

To prove this we work out addition and multiplication tables:

+	0	1	ζ	$1+\zeta$
0	0	1	ζ	$1+\zeta$
1	1	0	$1+\zeta$	ζ
ζ	ζ	$1+\zeta$	0	1
$1+\zeta$	$1+\zeta$	ζ	1	0

	0	1	ζ	$1+\zeta$
0	0	0	0	0
1	0	1	ζ	$1+\zeta$
ζ	0	ζ	$1+\zeta$	1
$1+\zeta$	0	$1+\zeta$	1	ζ

(The sort of calculation needed in the second table runs like this: $\zeta(1+\zeta) = \zeta + \zeta^2 = \zeta + \zeta + 1 = 1$.)

It follows that $\mathbf{Z}_2(\zeta)$ is a field with four elements. Now f splits over $\mathbf{Z}_2(\zeta)$:

$$t^2 + t + 1 = (t - \zeta)(t - 1 - \zeta)$$

but over no smaller field. Hence $\mathbf{Z}_2(\zeta)$ is a splitting field for f over $\mathbf{Z}_2$.

Normality

The idea of a normal extension was explicitly recognized by Galois (but as always in terms of polynomials over **C**). In the modern treatment it takes the following form:

Definition. An extension $L:K$ is *normal* if every irreducible polynomial f over K which has at least one zero in L splits in L.

This is a Trade Union definition – 'One out, all out!'

For example, **C**:**R** is normal since every polynomial (irreducible or not) splits in **C**. On the other hand we can find extensions which are not normal. Let α be the real cube root of 2 and consider $\mathbf{Q}(\alpha):\mathbf{Q}$. The irreducible polynomial $t^3 - 2$ has a zero, namely α, in $\mathbf{Q}(\alpha)$, but it does not split in $\mathbf{Q}(\alpha)$. If it did, then there would be three real cube roots of 2, not all equal. And this is absurd.

Compare these with the examples of Galois groups given in the preceding chapter. The normal extension **C**:**R** has a well-behaved Galois group, in the sense that the Galois correspondence is a bijection. The non-normal extension $\mathbf{Q}(\alpha):\mathbf{Q}$ has a badly behaved Galois group. Although this is not the whole story, it illustrates the importance of normality.

There is a close connection between normal extensions and splitting fields which provides a wide range of normal extensions:

8.4 Theorem. *An extension $L:K$ is normal and finite if and only if L is a splitting field for some polynomial over K.*

Proof. Suppose $L:K$ is normal and finite. By 4.4 $L = K(\alpha_1, \cdots, \alpha_s)$ for certain α_i algebraic over K. Let m_i be the minimum polynomial of α_i over K and let $f = m_1 \cdots m_s$. Each m_i is irreducible over K and has a zero $\alpha_i \in L$, so by normality each m_i splits over L. Hence f splits over L. Since L is generated by K and the zeros of f it is a splitting field for f over K.

To prove the converse, suppose that L is a splitting field for some polynomial g over K. The extension $L:K$ is then obviously finite; we must show it is normal. To do this we must take an irreducible polynomial f over K with a zero in L and show that it splits in L. Let $M \supseteq L$ be a splitting field for fg over K. Suppose that θ_1 and θ_2 are zeros of f in M. We claim that

$$[L(\theta_1):L] = [L(\theta_2):L].$$

We look at several subfields of M, related according to the following diagram:

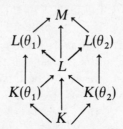

where the arrows denote inclusions. The claim will follow from a simple computation of degrees. For $j = 1$ or 2 we have

$$[L(\theta_j):L][L:K] = [L(\theta_j):K] = [L(\theta_j):K(\theta_j)][K(\theta_j):K].$$

$$(1)$$

Now by 4.3 $[K(\theta_1):K] = [K(\theta_2):K]$. Clearly $L(\theta_j)$ is a splitting field for g over $K(\theta_j)$, and by 3.8 $K(\theta_1)$ is isomorphic to $K(\theta_2)$. Therefore by 8.3 the extensions $L(\theta_j):K(\theta_j)$ are isomorphic for $j = 1, 2$; and hence have the same degree. Substituting in (1) and cancelling we find that

$$[L(\theta_1):L] = [L(\theta_2):L]$$

as claimed.

But then, if $\theta_1 \in L$ we have $[L(\theta_1):L] = 1$, so that $[L(\theta_2):L] = 1$ and $\theta_2 \in L$ also. Hence $L:K$ is normal.

Separability

Galois did not *explicitly* recognize the concept of separability, since he worked only with the complex field (where, as we shall see, separability is automatic). The concept is implicit in several of his proofs, and must be invoked when studying fields of non-zero characteristic.

Definition. An irreducible polynomial f over a field K is *separable over* K if it has no multiple zeros in a splitting field.

This means that in any splitting field f takes the form

$$k(t-\sigma_1) \cdots (t-\sigma_n)$$

where the σ_i are all different.

Definition. An irreducible polynomial over a field K is *inseparable over* K if it is not separable over K.

Examples

1 A separable polynomial. The polynomial $t^4+t^3+t^2+t+1$ is separable over $\mathbf{Q}$, for it has a splitting field inside $\mathbf{C}$ where the zeros are $e^{2\pi i/5}$, $e^{4\pi i/5}$, $e^{6\pi i/5}$, $e^{8\pi i/5}$, which are all different.

2 An inseparable polynomial. Let $K_0 = \mathbf{Z}_p$ for prime p. Let $K = K_0(u)$ where u is transcendental over K_0, and let $f(t) = t^p - u \in K[t]$. Let Σ be a splitting field for f over K, and let τ be a zero of f in Σ. Then $\tau^p = u$. Now

$$(t-\tau)^p = t^p + \binom{p}{1} t^{p-1}(-\tau) + \cdots + (-\tau)^p$$

by the Binomial Theorem. But all the Binomial coefficients $\binom{p}{r}$ where $0 < r < p$ are divisible by p, since in the expression $\dfrac{p!}{r!(p-r)!}$ the factor p in the numerator does not occur in the denominator. And in K any multiple of p is 0. Therefore

$$(t-\tau)^p = t^p - \tau^p = t^p - u = f(t).$$

Thus if $\sigma^p - u = 0$ then $(\sigma - \tau)^p = 0$ so that $\sigma = \tau$; all the zeros of f in Σ are equal.

It remains to show that f is irreducible over K. Suppose that $f = gh$ where $g, h \in K[t]$, and g and h have lower degree than f. We must have $g(t) = (t - \tau)^s$ where $0 < s < p$ by uniqueness of factorization. Hence the constant coefficient τ^s of g lies in K. This implies that $\tau \in K$, for there exist integers a and b such that $as + bp = 1$, and since $\tau^{as+bp} \in K$ it follows that $\tau \in K$. Then we must have $\tau = v(u)/w(u)$ where $v, w \in K_0[u]$, so that

$$v(u)^p - u(w(u))^p = 0.$$

But the terms of highest degree cannot cancel. Hence f is irreducible.

By the preceding paragraph f is inseparable over K.

Formal differentiation

For polynomials over **R** there is a standard method for detecting multiple zeros by differentiation. This method generalizes to arbitrary fields, but we must define the derivative in a purely formal manner.

Definition. Suppose that

$$f(t) = a_0 + a_1 t + \cdots + a_n t^n \in K[t]$$

for a field K. Then the *formal derivative* of f is the polynomial

$$Df = a_1 + 2a_2 t + \cdots + na_n t^{n-1}.$$

For $K = $ **R** this is the usual derivative. There is in general no hope of considering Df as a 'rate of change' of f, but certain of the more useful properties of the derivative do carry over to D. In particular, for all polynomials f and g over K we have

$$D(f+g) = Df + Dg$$
$$D(fg) = Df.g + f.Dg$$

and if $\lambda \in K$ then

$$D(\lambda) = 0$$

whence

$$D(\lambda f) = \lambda . Df.$$

These properties of D allow us to give a criterion for the existence of multiple zeros without knowing what the zeros are.

8.5 Lemma. *A polynomial $f \neq 0$ over a field K has a multiple zero in a splitting field if and only if f and Df have a common factor of degree ≥ 1.*

Proof. Suppose f has a repeated zero in a splitting field Σ, so that over Σ

$$f(t) = (t-\alpha)^2 g(t)$$

where $\alpha \in \Sigma$. Then

$$Df = (t-\alpha)\{(t-\alpha)Dg + 2g\}$$

so that f and Df have a common factor $(t-\alpha)$ in $\Sigma[t]$. Hence f and Df have a common factor in $K[t]$, namely the minimum polynomial of α over K.

Now suppose that f has no repeated zeros. We show by induction on ∂f that f and Df are coprime in $\Sigma[t]$, hence also coprime in $K[t]$. If $\partial f = 1$ this is obvious. Otherwise $f(t) = (t-\alpha)g(t)$ where $(t-\alpha) \nmid g(t)$. Then

$$Df = (t-\alpha)Dg + g.$$

If a factor of g divides Df it must also divide Dg since it is not $t-\alpha$. But by induction g and Dg are coprime. Hence f and Df are coprime, as required.

We can now give necessary and sufficient conditions for separability of an irreducible polynomial.

8.6 Proposition. *If K is a field of characteristic 0 then every*

irreducible polynomial over K is separable over K. If K has
characteristic p > 0 then an irreducible polynomial f over
K is inseparable if and only if

$$f(t) = k_0 + k_1 t^p + \cdots + k_r t^{rp}$$

where $k_0, \cdots, k_r \in K$.

Proof. An irreducible polynomial f over K is inseparable
if and only if f and Df have a common factor of degree ≥ 1.
Since f is irreducible and Df has smaller degree than f,
we must have $Df = 0$. Thus if

$$f(t) = a_0 + \cdots + a_n t^n$$

we have $na_n = 0$ for all integers $n > 0$. For characteristic 0
this implies that $a_n = 0$ for all n. For characteristic $p > 0$ it
implies that $a_n = 0$ if p does not divide n. Let $k_i = a_{ip}$, and
the result follows.

Note: The condition on f for inseparability over fields of
characteristic p can be expressed by saying that $f(t) = g(t^p)$
for some polynomial g over K.

We now define three more uses of the word 'separable'.

Definition. An arbitrary polynomial over a field K is
separable over K if all its irreducible factors are separable
over K.

If $L:K$ is an extension then an algebraic element $\alpha \in L$
is *separable* over K if its minimum polynomial over K is
separable over K.

An algebraic extension $L:K$ is a *separable* extension if
every $\alpha \in L$ is separable over K.

Finally we show that for algebraic extensions separability
carries over to intermediate fields.

8.7 Lemma. *Let $L:K$ be a separable algebraic extension and
let M be an intermediate field. Then $M:K$ and $L:M$ are
separable.*

Proof. Clearly $M:K$ is separable. Let $\alpha \in L$, and let m_K and m_M be its minimum polynomials over K, M respectively. Now $m_M | m_K$ in $M[t]$. But α is separable over K so m_K is separable over K, hence m_M is separable over M. Therefore $L:M$ is a separable extension.

Exercises

8.1 Construct subfields of **C** which are splitting fields over **Q** for the polynomials $t^3 - 1$, $t^4 + 5t^2 + 6$, $t^6 - 8$.

8.2 Find the degrees of these fields as extensions of **Q**.

8.3 Construct a splitting field for $t^3 + 2t + 1$ over $\mathbf{Z}_3$.

8.4 Construct a splitting field for $t^3 + t^2 + t + 2$ over $\mathbf{Z}_3$. Is it isomorphic to that constructed in Exercise 8.3?

8.5 List all quadratic monic polynomials over $\mathbf{Z}_5$. Which are irreducible? Construct splitting fields for some of the irreducible ones. Are these fields isomorphic? How many elements do they have?

8.6 Show that we can extend the definition of the formal derivative to $K(t)$ by defining

$$D(f/g) = (f.Dg - g.Df)/f^2.$$

Verify the relevant properties of D.

8.7 Which of the polynomials $t^3 + 1$, $t^2 + 2t - 1$, $t^6 + t^5 + t^4 + t^3 + t^2 + t + 1$, $7t^5 + t - 1$ are separable, considered as polynomials over the fields **Q**, **C**, $\mathbf{Z}_2$, $\mathbf{Z}_3$, $\mathbf{Z}_5$, $\mathbf{Z}_7$, $\mathbf{Z}_{19}$?

8.8 Which of the following extensions are normal?
(a) $\mathbf{Q}(t):\mathbf{Q}$
(b) $\mathbf{Q}(\sqrt{-5}):\mathbf{Q}$
(c) $\mathbf{Q}(\alpha):\mathbf{Q}$ where α is the real 7th root of 5

(d) $\mathbf{Q}(\sqrt{5}, \alpha):\mathbf{Q}(\alpha)$, where α is as in (c)

(e) $\mathbf{R}(\sqrt{-7}):\mathbf{R}$.

8.9 Show that every extension of degree 2 is normal. Is this true for any degree > 2?

8.10 If Σ is a splitting field for f over K and $K \subseteq L \subseteq \Sigma$, show that Σ is a splitting field for f over L.

8.11 Let f be a polynomial of degree n over K, and let Σ be a splitting field for f over K. Show that $[\Sigma:K]$ divides $n!$.

8.12 Mark the following true or false.

(a) Every polynomial splits over some field.

(b) The polynomial $t^3 + 5$ is separable over $\mathbf{Z}_7$.

(c) Splitting fields are unique up to isomorphism.

(d) Every finite extension is normal.

(e) Every separable extension is normal.

(f) Every finite normal extension is a splitting field for some polynomial.

(g) $\mathbf{Q}(\sqrt{19}):\mathbf{Q}$ is a separable normal extension.

(h) $\mathbf{Q}(\sqrt{21}):\mathbf{Q}$ is a separable normal extension.

(i) A reducible polynomial cannot be separable.

(j) If $Df = 0$ then $f = 0$, for a polynomial f over a field.

CHAPTER NINE

Field degrees and group orders

In proving the Fundamental Theorem of Galois Theory in Chapter 11 we need to show that if H is a subgroup of the Galois group of a finite separable normal extension $L:K$ then $H^{\dagger *} = H$. Our method will be to show that H and $H^{\dagger *}$ are finite groups and have the same order. Since we already know that $H \subseteq H^{\dagger *}$ the two groups must be equal. (It is largely for this reason that we need to restrict our attention to *finite* extensions and *finite* groups. If an infinite set is contained in another of the same cardinality, they need not be equal – for example $\mathbf{Z} \subseteq \mathbf{Q}$ and both sets are countable.)

The object of this chapter is to perform part of the calculation of the order of $H^{\dagger *}$, namely: we find the degree $[H^{\dagger}:K]$ in terms of the order of H. In the next chapter we shall find the order of $H^{\dagger *}$ in terms of this degree; putting the pieces together gives the desired result.

Our first result is due to Dedekind, who was the first to make a systematic study of field automorphisms.

9.1 Lemma. (*Dedekind*). *If K and L are fields, then every set of distinct monomorphisms $K \to L$ is linearly independent over L.*

Proof. Let $\lambda_1, \cdots, \lambda_n$ be distinct monomorphisms $K \to L$. To say these are linearly independent over L is to say that there do not exist elements $a_1, \cdots, a_n \in L$ such that

$$a_1 \lambda_1(x) + \cdots + a_n \lambda_n(x) = 0 \qquad (1)$$

for all $x \in K$, unless all the a_i are 0.

Assume the contrary, so that (1) holds. Without loss of generality all the a_i are non-zero. Of all the equations of the form (1) which hold, with all $a_i \neq 0$, there must be at least one for which the number n of terms is least. We choose notation so that (1) is such an expression. Hence we may assume that *there does not exist an equation like* (1) *with fewer than n terms*. We shall then deduce a contradiction.

There exists $y \in K$ such that $\lambda_1(y) \neq \lambda_n(y)$, since $\lambda_1 \neq \lambda_n$. Therefore $y \neq 0$. Now (1) holds with yx in place of x, so

$$a_1 \lambda_1(yx) + \cdots + a_n \lambda_n(yx) = 0$$

for all $x \in K$, whence

$$a_1 \lambda_1(y)\lambda_1(x) + \cdots + a_n \lambda_n(y)\lambda_n(x) = 0 \qquad (2)$$

for all $x \in K$. If we multiply (1) by $\lambda_1(y)$ and subtract (2) we obtain

$$a_2(\lambda_2(x)\lambda_1(y) - \lambda_2(x)\lambda_2(y)) + \cdots +$$
$$+ a_n(\lambda_n(x)\lambda_1(y) - \lambda_n(x)\lambda_n(y)) = 0.$$

The coefficient of $\lambda_n(x)$ is $a_n(\lambda_1(y) - \lambda_n(y)) \neq 0$, so we have an equation of form (1) with fewer terms. This contradicts our italicized assumption above.

Consequently no equation of the form (1) exists and the monomorphisms are linearly independent.

For our next result we need two lemmas. The first is a standard theorem of linear algebra which we quote without proof.

9.2 Lemma. *A system of m homogeneous equations*

$$a_{m1}x_1 + \cdots + a_{mn}x_n = 0$$

in n unknowns $x_1, \cdots, x_n$ with coefficients a_{ij} taken from a field has a solution in which the x_i are not all zero if $n > m$.

This theorem is proved in most first year undergraduate linear algebra courses, and can be found in any text of linear algebra (e.g. Halmos [11] p. 91).

The second lemma states a useful general principle.

9.3 Lemma. *If G is a group whose distinct elements are $g_1, \cdots, g_n$, and if $g \in G$, then as j varies from 1 to n the elements gg_j run through the whole of G, each element of G occurring precisely once.*

Proof. If $h \in G$ then $g^{-1}h = g_j$ for some j and $h = gg_j$. If $gg_i = gg_j$ then $g_i = g^{-1}gg_i = g^{-1}gg_j = g_j$.

We also require some notation.

Notation. The cardinality of a set S will be denoted by $|S|$. Thus if G is a group then $|G|$ is the order of G.

We now come to the main theorem of this chapter.

9.4 Theorem. *Let G be a finite subgroup of the group of automorphisms of a field K, and let K_0 be the fixed field of G. Then*

$$[K:K_0] = |G|.$$

Proof. Let $n = |G|$, and suppose that the elements of G are $g_1, \cdots, g_n$, where $g_1 = 1$.
(1) Suppose that $[K:K_0] = m < n$. Let $\{x_1, \cdots, x_m\}$ be a basis for K over K_0. By 9.2 we can find $y_1, \cdots, y_n \in K$, not all zero, such that

$$g_1(x_j)y_1 + \cdots + g_n(x_j)y_n = 0 \qquad (3)$$

for $j = 1, \cdots, m$. Let a be any element of K. Then

$$a = \alpha_1 x_1 + \cdots + \alpha_m x_m$$

where $\alpha_1, \cdots, \alpha_m \in K_0$. Hence

$$g_1(a)y_1 + \cdots + g_n(a)y_n = g_1(\sum_l \alpha_l x_l)y_1 + \cdots + g_n(\sum_l \alpha_l x_l)y_n$$

$$= \sum_l \alpha_l(g_1(x_l)y_1 + \cdots + g_n(x_l)y_n)$$

$$= 0$$

using (3). Hence the distinct monomorphisms $g_1, \cdots, g_n$ are linearly dependent, contrary to 9.1. Therefore $m \geq n$.

(2) Next suppose that $[K:K_0] > n$. Then there exists a set of $n+1$ elements of K linearly independent over K_0; let such a set be $\{x_1, \cdots, x_{n+1}\}$. By 9.2 there exist $y_1, \cdots, y_{n+1} \in K$, not all zero, such that for $j = 1, \cdots, n$

$$g_j(x_1)y_1 + \cdots + g_j(x_{n+1})y_{n+1} = 0. \tag{4}$$

We shall subject this to a combinatorial attack, similar to that used in proving 9.1. Choose $y_1, \cdots, y_{n+1}$ so that as few as possible are non-zero, and renumber so that

$$y_1, \cdots, y_r \neq 0, \quad y_{r+1}, \cdots, y_{n+1} = 0.$$

Equation (4) now becomes

$$g_j(x_1)y_1 + \cdots + g_j(x_r)y_r = 0. \tag{5}$$

Let $g \in G$, and operate on (5) with g. This gives a system of equations

$$gg_j(x_1)g(y_1) + \cdots + gg_j(x_r)g(y_r) = 0.$$

By 9.3 as j varies this system of equations is equivalent to the system

$$g_j(x_1)g(y_1) + \cdots + g_j(x_r)g(y_r) = 0. \tag{6}$$

If we multiply the equations (5) by $g(y_1)$ and (6) by y_1, and subtract, we obtain

$$g_j(x_2)(y_2 g(y_1) - g(y_2)y_1) + \cdots + g_j(x_r)(y_r g(y_1) - g(y_1)y_r) = 0.$$

This is a system of equations like (5) but with fewer terms, which gives a contradiction unless all the coefficients

$$y_i g(y_1) - y_1 g(y_i)$$

are zero. If this happens then

$$y_i y_1^{-1} = g(y_i y_1^{-1})$$

for all $g \in G$, so that $y_i y_1^{-1} \in K_0$. Thus there exist $z_1, \cdots, z_r \in K_0$ and an element $k \in K$ such that $y_i = kz_i$ for all i. Then equation (5), with $j = 1$, becomes

$$x_1 kz_1 + \cdots + x_r kz_r = 0$$

and since $k \neq 0$ we may divide by k, which shows that the x_i are linearly dependent over K_0. This is a contradiction. Therefore $[K:K_0] \leq n$, and then the first part of the proof shows that $[K:K_0] = n = |G|$ as required.

9.5 Corollary. *If G is the Galois group of the finite extension $L:K$ and H is a finite subgroup of G then*

$$[H^\dagger:K] = [L:K]/|H|.$$

Proof. $[H^\dagger:K] = [L:K]/[L:H^\dagger] = [L:K]/|H|$ by 9.4.

Examples. We illustrate Theorem 9.4 by two examples, one simple, the other more intricate.
1 Let G be the group of automorphisms of $\mathbf{C}$ consisting of the identity and complex conjugation. The fixed field of G is $\mathbf{R}$, for if $x - iy = x + iy$ ($x, y \in \mathbf{R}$) then $y = 0$, and conversely. Hence $[\mathbf{C}:\mathbf{R}] = |G| = 2$, a conclusion which is manifestly correct.
2 Let $K = \mathbf{Q}(\omega)$ where $\omega = e^{2\pi i/5} \in \mathbf{C}$. Now $\omega^5 = 1$ and $\mathbf{Q}(\omega)$ consists of all elements

$$p + q\omega + r\omega^2 + s\omega^3 + t\omega^4 \tag{7}$$

where $p, q, r, s, t \in \mathbf{Q}$. The Galois group of $\mathbf{Q}(\omega):\mathbf{Q}$ is easy to find, for if α is a $\mathbf{Q}$-automorphism of $\mathbf{Q}(\omega)$ then $(\alpha(\omega))^5 = \alpha(\omega^5) = \alpha(1) = 1$, so that $\alpha(\omega) = \omega$, ω^2, ω^3, or ω^4. This gives 4 candidates for $\mathbf{Q}$-automorphisms:

$\alpha_1: p + q\omega + r\omega^2 + s\omega^3 + t\omega^4 \rightarrow p + q\omega + r\omega^2 + s\omega^3 + t\omega^4$

$\alpha_2: \qquad\qquad\qquad\qquad \rightarrow p + s\omega + q\omega^2 + t\omega^3 + r\omega^4$

$\alpha_3: \qquad\qquad\qquad\qquad \rightarrow p + r\omega + t\omega^2 + q\omega^3 + s\omega^4$

$\alpha_4: \qquad\qquad\qquad\qquad \rightarrow p + t\omega + s\omega^2 + r\omega^3 + q\omega^4.$

It is easy to check that these are all $\mathbf{Q}$-automorphisms.
Hence the Galois group of $\mathbf{Q}(\omega):\mathbf{Q}$ has order 4. The fixed
field of this Galois group is easy to compute: it turns out
to be $\mathbf{Q}$. Therefore by 9.4 we should have $[\mathbf{Q}(\omega):\mathbf{Q}] = 4$.
At first sight this might seem wrong, for (7) expresses each
element in terms of 5 basis elements; the degree should be
5. In support of this contention, ω is a zero of $t^5 - 1$. The
astute reader will already have seen the source of this
dilemma: $t^5 - 1$ is *not* the minimum polynomial of ω over
$\mathbf{Q}$, since it is reducible. The minimum polynomial is in fact

$$t^4 + t^3 + t^2 + t + 1$$

which has degree 4. Equation (7) holds, but the elements of
the supposed 'basis' are linearly dependent:

$$\omega^4 + \omega^3 + \omega^2 + \omega + 1 = 0.$$

Hence every element of $\mathbf{Q}(\omega)$ can be expressed uniquely in
the form

$$p + q\omega + r\omega^2 + s\omega^3$$

where $p, q, r, s \in \mathbf{Q}$. We did not use this expression because
it lacks symmetry and makes the computations harder and
formless.

Exercises

9.1 Check theorem 9.4 for the extensions of Exercise 7.1
and their Galois groups.

9.2 Find the fixed field of the subgroup $\{\alpha_1, \alpha_4\}$ for the
second of the above examples. Check that 9.4 holds.

9.3 Parallel the argument of Example 2 above when
$\omega = e^{2\pi i/7}$.

9.4 Find all monomorphisms $\mathbf{Q} \to \mathbf{C}$.

9.5 Find all ring monomorphisms $\mathbf{Z} \to \mathbf{Q}$.

9.6 Show that any ring monomorphism $\mathbf{Z} \to K$, where K is a field, extends to a unique field monomorphism $\mathbf{Q} \to K$. Does this still hold good if K is not a field?

9.7 Mark the following true or false.
  (a) If $S \subseteq T$ is a finite set and $|S| = |T|$ then $S = T$.
  (b) The same is true of infinite sets.
  (c) There is only one monomorphism $\mathbf{Q} \to \mathbf{Q}$.
  (d) There are precisely p automorphisms of $\mathbf{Z}_p$.
  (e) If K and L are fields there exists at least one monomorphism $K \to L$.
  (f) If K and L are fields and there is a monomorphism $K \to L$ then K and L have the same characteristic.
  (g) If K and L are fields of equal characteristic there exists a monomorphism $K \to L$.
  (h) If K is a field of characteristic p there exists a unique monomorphism $\mathbf{Z}_p \to K$.
  (i) Distinct automorphisms of a field K are linearly independent over K.
  (j) Linearly independent monomorphisms are distinct.

CHAPTER TEN

Monomorphisms, automorphisms, and normal closures

The theme of this chapter is the construction of automorphisms to given specifications. We begin with a generalization of a K-automorphism, known as a K-monomorphism. For normal extensions we shall use K-monomorphisms to build up K-automorphisms. Using this we can calculate the order of the Galois group of any finite separable normal extension, which combines with the result of Chapter 9 to give a crucial part of the Fundamental Theorem of Chapter 11.

We also introduce the concept of a normal closure of a finite extension. This useful device enables us to steer around some of the technical obstructions caused by non-normal extensions.

K-monomorphisms

We begin by generalizing the idea of a K-automorphism of a field L.

Definition. Suppose that K is a subfield of each of the fields M and L. Then a K-*monomorphism* of M into L is a map $\phi : M \to L$ which is a field monomorphism such that $\phi(k) = k$ for every $k \in K$.

In general if $K \subseteq M \subseteq L$ then any K-automorphism of L restricts to a K-monomorphism $M \rightarrow L$. We are particularly interested in when this process can be reversed.

10.1 Theorem. *Suppose that $L:K$ is a finite normal extension and $K \subseteq M \subseteq L$. Let τ be any K-monomorphism $M \rightarrow L$. Then there exists a K-automorphism σ of L such that $\sigma|_M = \tau$.*

Proof. By 8.4 L is a splitting field over K of some polynomial f over K, hence it is simultaneously a splitting field over M for f and over $\tau(M)$ for f. But $\tau|_K$ is the identity so that $\tau(f) = f$. We have the diagram

$$
\begin{array}{ccc}
M & \rightarrow & L \\
\tau \downarrow & & \downarrow \sigma \\
\tau(M) & \rightarrow & L
\end{array}
$$

so by Theorem 8.3 there is an isomorphism $\sigma:L \rightarrow L$ such that $\sigma|_M = \tau$. Therefore σ is an automorphism of L, and since $\sigma|_K = \tau|_K$ is the identity, σ is a K-automorphism of L.

This result can be used to construct K-automorphisms as follows:

10.2 Proposition. *Suppose $L:K$ is a finite normal extension and α, β are zeros in L of the irreducible polynomial p over K. Then there exists a K-automorphism σ of L such that $\sigma(\alpha) = \beta$.*

Proof. By 3.8 there is an isomorphism $\tau:K(\alpha) \rightarrow K(\beta)$ such that $\tau|_K$ is the identity and $\tau(\alpha) = \beta$. By 10.1 τ extends to a K-automorphism σ of L.

Normal closures

When extensions are not normal, we can try to recover normality by making the extensions larger.

Definition. Let L be an algebraic extension of K. A *normal*

closure of $L:K$ is an extension N of L such that
(1) $N:K$ is normal,
(2) If $L \subseteq M \subseteq N$ and $M:K$ is normal then $M = N$.
Thus N is the smallest extension of L which is normal over K.

The next theorem assures us of a sufficient supply of normal closures.

10.3 Theorem. *If $L:K$ is a finite extension then there exists a normal closure N which is a finite extension of K. If M is another normal closure then the extensions $M:K$ and $N:K$ are isomorphic.*

Proof. Let $x_1, \cdots, x_r$ be a basis for L over K, and let m_i be the minimum polynomial of x_i over K. Let N be a splitting field for $f = m_1 m_2 \cdots m_r$ over L. Then N is also a splitting field for f over K so that $N:K$ is normal and finite by 8.4. Suppose that $L \subseteq P \subseteq N$ where $P:K$ is normal. Each polynomial m_i has a zero $x_i \in P$, so by normality f splits in P. Since N is a splitting field for f we have $P = N$. Therefore N is a normal closure.

Now suppose that M and N are both normal closures. The above polynomial f splits in M and in N, so each of M and N contain a splitting field for f over K. These splitting fields contain L and are normal over K, so must be equal to M and N respectively. The uniqueness of splitting fields (8.3) now says that $M:K$ and $N:K$ are isomorphic.

For example, consider $\mathbf{Q}(c):\mathbf{Q}$ where c is the real cube root of 2. This extension is not normal, as we have seen. If we let K be a splitting field for $t^3 - 2$ over $\mathbf{Q}$, contained in C, then $K = \mathbf{Q}(c, c\omega, c\omega^2)$ where $\omega = \dfrac{-1 + i\sqrt{3}}{2}$ is a complex cube root of unity. Now K is a normal closure for $\mathbf{Q}(c):\mathbf{Q}$. We obtain a normal closure by adjoining all the 'missing' zeros.

Normal closures enable us to place restrictions on the image of a monomorphism.

10.4 Lemma. *Suppose that $K \subseteq L \subseteq N \subseteq M$ where $L:K$ is finite and N is a normal closure of $L:K$. Let τ be any K-monomorphism $L \to M$. Then $\tau(L) \subseteq N$.*

Proof. Let $\alpha \in L$. Then α has minimum polynomial m over K. Then $0 = m(\alpha) = \tau(m(\alpha)) = m(\tau(\alpha))$ so that $\tau(\alpha)$ is a zero of m, and lies in N since $N:K$ is normal. Therefore $\tau(L) \subseteq N$.

This result often allows us to restrict our attention to a normal closure of a given extension when discussing monomorphisms. The next theorem provides a sort of converse.

10.5 Theorem. *For a finite extension $L:K$ the following are equivalent:*

(1) $L:K$ is normal.

(2) There exists a normal extension N of K containing L such that every K-monomorphism $\tau:L \to N$ is a K-automorphism of L.

(3) For every normal extension M of K containing L every K-monomorphism $\tau:L \to M$ is a K-automorphism of L.

Proof. We show that $(1) \Rightarrow (3) \Rightarrow (2) \Rightarrow (1)$.

$(1) \Rightarrow (3)$. If $L:K$ is normal then L is a normal closure of $L:K$ so by 10.4 $\tau(L) \subseteq L$. But τ is a K-linear map defined on the finite-dimensional vector space L over K, and is a monomorphism, so that $\tau(L)$ has the same dimension as L, whence $\tau(L) = L$ and τ is a K-automorphism of L.

$(3) \Rightarrow (2)$. Let N be a normal closure for $L:K$. Then N exists by 10.3 and has the requisite properties by (3).

$(2) \Rightarrow (1)$. Suppose f is any irreducible polynomial over K with a zero $\alpha \in L$. Then f splits over N by normality, and if β is any zero of f in N there exists an automorphism σ of N such that $\sigma(\alpha) = \beta$ by 10.2. By hypothesis σ is a K-automorphism of L, so that $\beta = \sigma(\alpha) \in \sigma(L) = L$. Therefore f splits over L and $L:K$ is normal.

Our next result is of a more computational nature.

10.6 Theorem. *Suppose $L:K$ is a finite separable extension* *of degree n. Then there are precisely n distinct K-monomorphisms of L into a normal closure N (and hence into any given normal extension M of K containing L).*

Proof. We use induction on $[L:K]$. If $[L:K] = 1$ the result is clear. Suppose that $[L:K] = k > 1$. Let $\alpha \in L\backslash K$ with minimum polynomial m over K. Then $\partial m = [K(\alpha):K] = r > 1$. Now m is a separable irreducible polynomial with one zero in the normal extension N, so m splits in N and its zeros $\alpha_1, \cdots, \alpha_r$ are distinct. By induction there are precisely s distinct $K(\alpha)$-monomorphisms $\rho_1, \cdots, \rho_s : L \to N$, where $s = [L:K(\alpha)] = k/r$. By 10.2 there are r distinct K-automorphisms $\tau_1, \cdots, \tau_r$ of N, such that $\tau_i(\alpha) = \alpha_i$. The maps

$$\phi_{ij} = \tau_i \rho_j$$

give $rs = k$ distinct K-monomorphisms $L \to N$. We show that these exhaust the K-monomorphisms $L \to N$.

Let $\tau : L \to N$ be a K-monomorphism. Then $\tau(\alpha)$ is a zero of m in N, so that $\tau(\alpha) = \alpha_i$ for some i. The map $\phi = \tau_i^{-1}\tau$ is a $K(\alpha)$-monomorphism $L \to N$, so by induction $\phi = \rho_j$ for some j. Hence $\tau = \tau_i \rho_j = \phi_{ij}$ and the theorem is proved.

We can now calculate the order of the Galois group of a finite separable normal extension.

10.7 Corollary. *If $L:K$ is a finite separable normal extension of degree n then there are precisely n distinct K-automorphisms of L, so that $|\Gamma(L:K)| = n$.*

Proof. Use 10.6 and 10.5.

From this we easily deduce the important:

10.8 Theorem. *Let $L:K$ be a finite extension with Galois group G. If $L:K$ is normal and separable then K is the fixed field of G.*

Proof. Let K_0 be the fixed field of G, and let $[L:K] = n$. By 10.7 $|G| = n$. By 9.4 $[L:K_0] = n$. Since $K \subseteq K_0$ we must have $K = K_0$.

There is a converse to this result which shows why we must consider separable normal extensions in order to make the Galois correspondence a bijection. Before we can prove the converse we need a theorem whose statement and proof closely resemble those of 10.6.

10.9 Theorem. *Suppose that $K \subseteq L \subseteq M$, that $M:K$ is finite, and that $[L:K] = n$. Then there are at most n K-monomorphisms $L \to M$.*

Proof. Let N be a normal closure of $M:K$. Then $N:K$ is finite by 10.3, and every K-monomorphism $L \to M$ is also a K-monomorphism $L \to N$. Hence we may assume that M is a normal extension of K by replacing M by N. We now argue by induction on $[L:K]$ as in the proof of 10.6, except that we can now only deduce that there are s' $K(\alpha)$-monomorphisms $L \to N$, where $s' \le s$ (by induction) and there are r' distinct K-automorphisms of N, where $r' \le r$ (since the zeros of m in N need not be distinct). The rest of the argument goes through.

Note further that if $L:K$ is not separable then there are *fewer* than n K-monomorphisms $L \to M$, since r' is definitely less than r for some choice of α.

10.10 Theorem. *If $L:K$ is a finite extension with Galois group G and if K is the fixed field of G then $L:K$ is normal and separable.*

Proof. By 9.4 $[L:K] = |G| = n$, say. There are exactly n distinct K-monomorphisms $L \to L$, namely, the elements of the Galois group. But as noted in the proof of 10.9, if $L:K$ is not separable there are $< n$ K-monomorphisms $L \to L$. Therefore $L:K$ is separable.

We prove normality by using 10.5. Thus let N be a normal extension of K containing L and let τ be a K-monomorphism $L \to N$. Since every element of the Galois group of $L:K$ defines a K-monomorphism $L \to N$ there are n K-monomorphisms $L \to N$ which are automorphisms of L. But by 10.9 τ can take at most n values, so τ is one of these monomorphisms. Hence τ is an automorphism of L. By 10.5 $L:K$ is normal.

If the Galois correspondence is a bijection then K must be the fixed field of the Galois group of $L:K$, so by the above $L:K$ must be separable and normal. That these hypotheses are also sufficient to make the Galois correspondence bijective will be proved in the next chapter.

Exercises

10.1 Suppose that $L:K$ is finite. Show that every K-monomorphism $L \to L$ is an automorphism. Need this result hold if the extension is not finite?

10.2 Construct normal closures N for the following extensions.
(a) $\mathbf{Q}(\alpha):\mathbf{Q}$ where α is the real 5th root of 3.
(b) $\mathbf{Q}(\beta):\mathbf{Q}$ where β is the real 7th root of 2.
(c) $\mathbf{Q}(\sqrt{2}, \sqrt{3}):\mathbf{Q}$.
(d) $\mathbf{Q}(\alpha, \sqrt{2}):\mathbf{Q}$ where α is the real cube root of 2.
(e) $\mathbf{Q}(\gamma):\mathbf{Q}$ where γ is a zero of $t^3 - 3t^2 + 3$.

10.3 Find the Galois groups of the extensions (a), (b), (c), (d) in Question 10.2.

10.4 Find the Galois groups of the extensions $N:\mathbf{Q}$ for their normal closures N.

10.5 Show that Lemma 10.4 fails if we do not assume that $N:K$ is normal, but is true for any extension N of L such that $N:K$ is normal, rather than just for a normal closure.

10.6 Use 10.7 to find the order of the Galois group of $\mathbb{Q}(\sqrt{2}, \sqrt{3}, \sqrt{5}):\mathbb{Q}$.

10.7 Mark the following true or false.
  (a) Every K-monomorphism is a K-automorphism.
  (b) Every finite extension has a normal closure.
  (c) If $K \subseteq L$ and σ is a K-automorphism of L, then $\sigma|_K$ is a K-automorphism of K.
  (d) Every extension of a field of characteristic 0 is normal.
  (e) An extension having Galois group of order 1 is normal.
  (f) A finite separable normal extension has finite Galois group.
  (g) Every Galois group is abelian.
  (h) The Galois correspondence fails for non-normal extensions.
  (i) A finite separable normal extension of degree n has Galois group of order n.
  (j) The Galois group of a normal extension is cyclic.

CHAPTER ELEVEN

The Galois correspondence

We are at last in a position to establish the fundamental properties of the Galois correspondence between a field extension and its Galois group. Most of the work has already been done, and all that remains is to put the pieces together.

Let us recall a few points of notation from Chapter 7. Let $L:K$ be a field extension with Galois group G, which consists of all K-automorphisms of L. Let $\mathscr{F}$ be the set of intermediate fields M, and $\mathscr{G}$ the set of all subgroups H of G. We have defined two maps

$$*: \mathscr{F} \to \mathscr{G}$$

$$^\dagger: \mathscr{G} \to \mathscr{F}$$

as follows: if $M \in \mathscr{F}$ then M^* is the group of all M-automorphisms of L. If $H \in \mathscr{G}$ then $H^\dagger$ is the fixed field of H. We have observed that the maps $*$ and † reverse inclusions, that $M \subseteq M^{*\dagger}$, and $H \subseteq H^{\dagger *}$.

11.1 Theorem (*Fundamental Theorem of Galois Theory*). *If $L:K$ is a finite separable normal field extension of degree n, with Galois group G; and if $\mathscr{F}$, $\mathscr{G}$, $*$, † are defined as above, then:*
(1) *The Galois group G has order n.*

(2) *The maps * and † are mutual inverses and set up an order-reversing 1–1 correspondence between $\mathscr{F}$ and $\mathscr{G}$.*

(3) *If M is an intermediate field then*

$$[L:M] = |M^*|$$
$$[M:K] = |G|/|M^*|.$$

(4) *An intermediate field M is a normal extension of K if and only if M^* is a normal subgroup of G (in the usual sense of group theory).*

(5) *If an intermediate field M is a normal extension of K then the Galois group of M : K is isomorphic to the quotient group G/M^*.*

Proof. The first part is a restatement of 10.7. For the second part, we know by 8.7 that $L:M$ is separable, and from 8.4 $L:M$ is clearly normal. Therefore by 10.8 M is the fixed field of M^*. Therefore

$$M^{*\dagger} = M. \tag{1}$$

Now consider $H \in \mathscr{G}$. We know that $H \subseteq H^{\dagger*}$. Now $H^{\dagger*\dagger} = (H^{\dagger})^{*\dagger} = H^{\dagger}$ by equation (1) above. By 9.4 we have

$$|H| = [K:H^{\dagger}].$$

Therefore

$$|H| = [K:H^{\dagger*\dagger}]$$

and by 9.4 again we have

$$[K:H^{\dagger*\dagger}] = |H^{\dagger*}|$$

so that

$$|H| = |H^{\dagger*}|.$$

Since H and $H^{\dagger*}$ are finite groups and $H \subseteq H^{\dagger*}$ we must have

$$H = H^{\dagger*}.$$

The second part of the theorem follows at once.

For the third part we again note that $L:M$ is separable and normal. By 10.7 $[L:M] = |M^*|$, and the other equality follows immediately.

Note that $|G|/|M^*|$ is the *index* of M^* in G in the usual sense of group theory.

To prove the last two parts of the theorem we require a lemma.

11.2 Lemma. *Suppose that $L:K$ is a finite separable normal extension, M is an intermediate field, and τ is a K-automorphism of L. Then*

$$(\tau(M))^* = \tau M^* \tau^{-1}.$$

Proof. Let $M' = \tau(M)$, and take $\gamma \in M^*$, $x_1 \in M'$. Then $x_1 = \tau(x)$ from some $x \in M$. Then

$$(\tau\gamma\tau^{-1})(x_1) = \tau\gamma(x) = \tau(x) = x_1$$

so that

$$\tau M^* \tau^{-1} \subseteq M'^*.$$

Similarly

$$\tau^{-1} M'^* \tau \subseteq M^*,$$

so that

$$\tau M^* \tau^{-1} \supseteq M'^*,$$

and the lemma is proved.

We now prove the fourth part of the theorem. If $M:K$ is normal, let $\tau \in G$. Then $\tau|_M$ is a K-monomorphism $M \to L$, so is a K-automorphism of M by 10.5. Hence $\tau(M) = M$. By Lemma 11.2 we have $\tau M^* \tau^{-1} = M^*$, so that M^* is a normal subgroup of G.

Conversely, suppose that M^* is a normal subgroup of G. Let σ be any K-monomorphism $M \to L$. By 10.1 there is a K-automorphism τ of L such that $\tau|_M = \sigma$. Now $\tau M^* \tau^{-1} = M^*$ since M^* is a normal subgroup of G, so by 11.2 $(\tau(M))^* = M^*$. By part 2 of the theorem, this means that $\tau(M) = M$. Hence $\sigma(M) = M$, and σ is a K-automorphism of M. By 10.5 $M:K$ is normal.

Now we prove the final part of the theorem. Let G' be the Galois group of $M:K$. We can define a map $\phi:G \rightarrow G'$ by

$$\phi(\tau) = \tau|_M \quad (\tau \in G).$$

This is clearly a group homomorphism $G \rightarrow G'$, for by 10.5 $\tau|_M$ is a K-automorphism of M. By 10.1 ϕ is surjective. The kernel of ϕ is obviously M^*. So by standard group theory

$$G' = im(\phi) \cong G/\ker(\phi) = G/M^*.$$

Note how Theorem 9.4 is used in the proof of part 2 of the theorem; its use is crucial. Many of the most beautiful results in mathematics hang by equally slender threads.

The importance of the Fundamental Theorem derives from its potential as a tool rather than its intrinsic merit. It enables us to apply group theory to otherwise intractable problems about fields (or polynomials), and we shall spend most of the remaining chapters exploiting such applications. But before venturing further we shall consolidate our position by illustrating the whole theory with regard to a particular field extension and its Galois group. The next chapter is devoted to this end.

CHAPTER TWELVE

A specific example

The extension we shall discuss in this chapter is a favourite with writers on field theory because of its archetypal quality. A simpler example would be too small to illustrate the theory adequately, and anything more complicated would be unwieldy.

The discussion will be cut into small pieces to make it more easily digestible.

1 Let $f(t) = t^4 - 2$ over $\mathbf{Q}$, and let K be a splitting field for f such that $K \subseteq \mathbf{C}$. In $\mathbf{C}$ we can factorize f as follows:

$$f(t) = (t - \xi)(t + \xi)(t - i\xi)(t + i\xi)$$

where $\xi = \sqrt[4]{2}$ is real and positive. Clearly therefore $K = \mathbf{Q}(\xi, i)$. The characteristic is 0 and K is a splitting field, so that $K : \mathbf{Q}$ is finite, separable, and normal.

2 We shall find the degree of $K : \mathbf{Q}$. We have

$$[K : \mathbf{Q}] = [\mathbf{Q}(\xi, i) : \mathbf{Q}(\xi)][\mathbf{Q}(\xi) : \mathbf{Q}].$$

The minimum polynomial of i over $\mathbf{Q}(\xi)$ is $t^2 + 1$ since $i^2 + 1 = 0$ but $i \notin \mathbf{R} \supseteq \mathbf{Q}(\xi)$. So $[\mathbf{Q}(\xi, i) : \mathbf{Q}(\xi)] = 2$.

Now ξ is a zero of f over $\mathbf{Q}$, and f is irreducible by Eisenstein's criterion, 2.5. Hence f is the minimum polynomial of ξ over $\mathbf{Q}$, and $[\mathbf{Q}(\xi):\mathbf{Q}] = 4$. Therefore

$$[K:\mathbf{Q}] = 2.4 = 8.$$

3 We shall find the elements of the Galois group of $K:\mathbf{Q}$. By a direct check, or by several applications of 3.9, we see that there is a $\mathbf{Q}$-automorphism σ of K such that

$$\sigma(i) = i, \quad \sigma(\xi) = i\xi$$

and another, τ, such that

$$\tau(i) = -i, \quad \tau(\xi) = \xi.$$

Products of these yield 8 distinct $\mathbf{Q}$-automorphisms of K, as follows:

automorphism	effect on ξ	effect on i
1	ξ	i
σ	$i\xi$	i
σ^2	$-\xi$	i
σ^3	$-i\xi$	i
τ	ξ	$-i$
$\sigma\tau$	$i\xi$	$-i$
$\sigma^2\tau$	$-\xi$	$-i$
$\sigma^3\tau$	$-i\xi$	$-i$

Other products do not give new automorphisms, since $\sigma^4 = 1$, $\tau^2 = 1$, $\tau\sigma = \sigma^3\tau$, $\tau\sigma^2 = \sigma^2\tau$, $\tau\sigma^3 = \sigma\tau$. (The last two relations follow from the first three.)

Now any $\mathbf{Q}$-automorphism of K sends i to some zero of $t^2 + 1$, so $i \rightarrow \pm i$; similarly ξ is mapped to $\xi, i\xi, -\xi$, or $-i\xi$. All possible combinations of these (8 in number) appear in the above list, so these are precisely the $\mathbf{Q}$-automorphisms of K.

4 The abstract structure of the Galois group G can be found. From the generator-relation presentation

$$G = \langle \sigma, \tau : \sigma^4 = \tau^2 = 1, \tau\sigma = \sigma^3\tau \rangle$$

it follows that G is the dihedral group of order 8, which we shall write as $\mathbf{D}_8$.

5 It is an easy exercise to find the subgroups of G. If we let $\mathbf{C}_n$ denote the cyclic group of order n, and $\times$ the direct product, then the subgroups are as follows:

Order 8: G	$G \cong \mathbf{D}_8$
Order 4: $\{1, \sigma, \sigma^2, \sigma^3\}$	$S \cong \mathbf{C}_4$
$\{1, \sigma^2, \tau, \sigma^2\tau\}$	$T \cong \mathbf{C}_2 \times \mathbf{C}_2$
$\{1, \sigma^2, \sigma\tau, \sigma^3\tau\}$	$U \cong \mathbf{C}_2 \times \mathbf{C}_2$
Order 2: $\{1, \sigma^2\}$	$A \cong \mathbf{C}_2$
$\{1, \tau\}$	$B \cong \mathbf{C}_2$
$\{1, \sigma\tau\}$	$C \cong \mathbf{C}_2$
$\{1, \sigma^2\tau\}$	$D \cong \mathbf{C}_2$
$\{1, \sigma^3\tau\}$	$E \cong \mathbf{C}_2$
Order 1: $\{1\}$	$I \cong \mathbf{C}_1$

6 The inclusion relations between the subgroups of G can be summed up by the following *lattice diagram*:

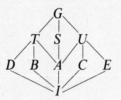

(Here $X \subseteq Y$ if there is a sequence of upward-sloping lines from X to Y.)

7 Under the Galois correspondence we obtain the intermediate fields. Since the correspondence reverses inclusions this gives a lattice diagram of fields as follows:

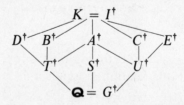

8 We now describe the elements of these intermediate fields.

There are three obvious subfields of K of degree 2 over $\mathbf{Q}$, namely $\mathbf{Q}(i)$, $\mathbf{Q}(\sqrt{2})$, $\mathbf{Q}(i\sqrt{2})$. These are clearly the fixed fields $S^\dagger$, $T^\dagger$, and $U^\dagger$ (respectively). The other fixed fields are less obvious. To illustrate a possible approach we shall find $C^\dagger$.

Now any element of K can be expressed in the form

$$x = a_0 + a_1\xi + a_2\xi^2 + a_3\xi^3 + a_4 i + a_5 i\xi + a_6 i\xi^2 + a_7 i\xi^3$$

where $a_0, \cdots, a_7 \in \mathbf{Q}$. Then

$$\sigma\tau(x) = a_0 + a_1 i\xi - a_2\xi^2 - a_3 i\xi^3 - a_4 i + a_5(-i)i\xi - a_6 i(i\xi)^2 - $$
$$- a_7 i(i\xi)^3$$
$$= a_0 + a_5\xi - a_2\xi^2 - a_7\xi^3 - a_4 i + a_1 i\xi + a_6 i\xi^2 - a_3 i\xi^3.$$

Therefore x is fixed by $\sigma\tau$ (and hence by C) if and only if $a_0 = a_0, a_1 = a_5, a_2 = -a_2, a_3 = -a_7, a_4 = -a_4, a_5 = a_1,$ $a_6 = a_6, a_7 = -a_3$. Therefore a_0 and a_6 are arbitrary, $a_2 = 0 = a_4, a_1 = a_5$, and $a_3 = -a_7$. It follows that

$$x = a_0 + a_1(1+i)\xi + a_6 i\xi^2 + a_3(1-i)\xi^3$$

$$= a_0 + a_1\{(1+i)\xi\} + \frac{a_6}{2}\{(1+i)\xi\}^2 - \frac{a_3}{2}\{(1+i)\xi\}^3$$

which means that

$$C^\dagger = \mathbf{Q}((1+i)\xi).$$

Similarly we have

$$A^\dagger = \mathbf{Q}(i, \sqrt{2})$$
$$B^\dagger = \mathbf{Q}(\xi)$$
$$D^\dagger = \mathbf{Q}(i\xi)$$
$$E^\dagger = \mathbf{Q}((1-i)\xi).$$

It is now easy to verify the inclusion relations determined by the lattice diagram of Section 7.

9 It is left to the reader (be he so foolhardy) to check by hand that these are indeed the only intermediate fields.

10 The normal subgroups of G are G, S, T, U, A, I. By the theory, $G^\dagger$, $S^\dagger$, $T^\dagger$, $U^\dagger$, $A^\dagger$, $I^\dagger$ should be the only normal extensions of $\mathbf{Q}$ which are contained in K. Since these are all splitting fields over $\mathbf{Q}$, for the polynomials $t, t^2 + 1, t^2 - 2,$ $t^2 + 2, t^4 - t^2 - 2, t^4 - 2$ (respectively) they are normal extensions of $\mathbf{Q}$. On the other hand $B^\dagger : \mathbf{Q}$ is not normal, since $t^4 - 2$ has a zero, namely ξ, in $B^\dagger$ but does not split in $B^\dagger$. Similarly $C^\dagger$, $D^\dagger$, $E^\dagger$ are not normal extensions of $\mathbf{Q}$.

11 According to the theory the Galois group of $A^\dagger : \mathbf{Q}$ is isomorphic to G/A. Now G/A is isomorphic to $\mathbf{C}_2 \times \mathbf{C}_2$. We calculate directly the Galois group of $A^\dagger : \mathbf{Q}$. Since $A^\dagger = \mathbf{Q}(i, \sqrt{2})$ there are 4 $\mathbf{Q}$-automorphisms:

automorphism	effect on i	effect on $\sqrt{2}$
1	i	$\sqrt{2}$
α	i	$-\sqrt{2}$
β	$-i$	$\sqrt{2}$
$\alpha\beta$	$-i$	$-\sqrt{2}$

and since $\alpha^2 = \beta^2 = 1$ and $\alpha\beta = \beta\alpha$, this group is $\mathbf{C}_2 \times \mathbf{C}_2$ as expected.

12 Note that the lattice diagrams for $\mathscr{F}$ and $\mathscr{G}$ do not look the same unless one is turned upside-down. Hence there does not exist a correspondence like the Galois correspondence but preserving inclusion relations. It may seem a little odd at first that the Galois correspondence *reverses* inclusions, but in fact it is entirely natural, and quite as useful a property as preservation of inclusions.

13 We should end this chapter by pointing out that it is in general a difficult problem to compute the Galois group of a given field extension, particularly when there is no explicit representation for the elements of the large field.

Exercises

12.1 Find the Galois groups of the following extensions:
    (a) $\mathbf{Q}(\sqrt{2}, \sqrt{5}) : \mathbf{Q}$

(b) $\mathbf{Q}(\alpha):\mathbf{Q}$ where $\alpha = e^{2\pi i/3}$

(c) $K:\mathbf{Q}$ where K is a splitting field over $\mathbf{Q}$ for $t^4 - 3t^2 + 4$.

12.2 Find the subgroups of these Galois groups.

12.3 Find the corresponding fixed fields.

12.4 Find the normal subgroups of the above Galois groups.

12.5 Check that the corresponding extensions are normal.

12.6 Verify that the Galois groups of these normal extensions are the relevant quotient groups.

CHAPTER THIRTEEN

Some group theory

In order to apply the Galois correspondence we need to have at our fingertips a number of group-theoretical concepts and theorems. We shall assume that the reader is familiar with elementary group theory: subgroups, normal subgroups, quotient groups, conjugates, permutations (up to cycle decomposition), and the standard isomorphism theorems. The relevant theory (along with most of the material in this chapter) can be found in any textbook of group theory, for example Ledermann [19] or MacDonald [20].

The first section of this chapter is devoted to defining soluble groups and elucidating some basic properties. These groups are of cardinal importance for the theory of solution of equations by radicals. The second section is a discussion of simple groups, the main target being a proof of the simplicity of the alternating group of degree greater than 4. The third section provides an introduction to the theory of p-groups and to Sylow theory.

Soluble groups

Soluble groups were first defined and studied by Galois in his work on the solution of equations by radicals. They have since proved extremely important in many branches of mathematics (particularly group theory!).

In the following definition, and thereafter, the notation $H \lhd G$ will mean that H is a normal subgroup of the group G.

Definition. A group G is *soluble* if it has a finite series of subgroups

$$1 = G_0 \subseteq G_1 \subseteq \cdots \subseteq G_n = G \qquad (1)$$

such that

(1) $G_i \lhd G_{i+1}$ for $i = 0, \cdots, n-1$,
(2) G_{i+1}/G_i is abelian for $i = 0, \cdots, n-1$.

Note that condition (1) does *not* imply that $G_i \lhd G$, since $G_i \lhd G_{i+1} \lhd G_{i+2}$ does not imply $G_i \lhd G_{i+2}$.

Examples
1 Every abelian group G is soluble, with series $1 \subseteq G$.
2 The symmetric group $\mathbf{S}_3$ of degree 3 is soluble, since it has a normal subgroup of order 3 generated by the cycle (123) whose quotient is cyclic of order 2.
3 The dihedral group $\mathbf{D}_8$ of order 8 is soluble. In the notation of Chapter 12 it has a normal subgroup S of order 4 and whose quotient has order 2. And S is abelian.
4 The symmetric group $\mathbf{S}_4$ of degree 4 is soluble, having a series

$$1 \lhd \mathbf{V} \lhd \mathbf{A}_4 \lhd \mathbf{S}_4$$

where $\mathbf{A}_4$ is the alternating group of order 12, and $\mathbf{V}$ consists of the permutations 1, (12)(34), (13)(24), (14)(23) and is a direct product of two cyclic groups of order 2. (The symbol $\mathbf{V}$ comes from Klein's term *Vierergruppe*, or *four-group*.) The quotient groups are

$$\mathbf{V}/1 \cong \mathbf{V} \quad \text{abelian of order 4}$$

$$\mathbf{A}_4/\mathbf{V} \cong \mathbf{C}_3 \quad \text{abelian of order 3}$$

$$\mathbf{S}_4/\mathbf{A}_4 \cong \mathbf{C}_2 \quad \text{abelian of order 2.}$$

5 The symmetric group $\mathbf{S}_5$ of degree 5 is *not* soluble, but we must defer the proof until Corollary 13.5.

We recall the following *Isomorphism Theorems*:

13.1 Lemma. *Let G, H, and A be groups.*

(1) *If $H \lhd G$ and $A \subseteq G$ then $H \cap A \lhd A$ and*

$$\frac{A}{H \cap A} \cong \frac{HA}{H}.$$

(2) *If $H \lhd G$ and $H \subseteq A \lhd G$ then $H \lhd A$, $A/H \lhd G/H$, and*

$$\frac{G/H}{A/H} \cong \frac{G}{A}.$$

(Parts (1) and (2) are respectively the *first* and *second*‡ isomorphism theorems.)

Judicious use of these isomorphism theorems allows us to prove that soluble groups persist in being soluble even when subjected to quite drastic treatment.

13.2 Theorem. *Let G be a group, H a subgroup of G, and N a normal subgroup of G.*
(1) *If G is soluble then H is soluble.*
(2) *If G is soluble then G/N is soluble.*
(3) *If N and G/N are soluble then G is soluble.*

Proof. (1) Let

$$1 = G_0 \lhd G_1 \lhd \cdots \lhd G_r = G$$

be a series for G with abelian factors G_{i+1}/G_i. Let $H_i = G_i \cap H$. Then H has a series

$$1 = H_0 \lhd \cdots \lhd H_r = H.$$

We show the factors are abelian. Now

$$\frac{H_{i+1}}{H_i} = \frac{G_{i+1} \cap H}{G_i \cap H} = \frac{G_{i+1} \cap H}{G_i \cap (G_{i+1} \cap H)} \cong \frac{G_i(G_{i+i} \cap H)}{G_i}$$

‡ Or occasionally the *second* and *third* isomorphism theorems. . . .

by the first isomorphism theorem. But this latter group is a subgroup of G_{i+1}/G_i which is abelian. Hence H_{i+1}/H_i is abelian and H is soluble.

(2) Define G_i as before. Then G/N has a series

$$N/N = G_0N/N \lhd G_1N/N \lhd \cdots \lhd G_rN/N = G/N.$$

A typical quotient is

$$\frac{G_{i+1}N/N}{G_iN/N}$$

which by the second isomorphism theorem is isomorphic to

$$\frac{G_{i+1}N}{G_iN} = \frac{G_{i+1}(G_iN)}{G_iN} \cong \frac{G_{i+1}}{G_{i+1} \cap (G_iN)} \cong \frac{G_{i+1}/G_i}{(G_{i+1} \cap (G_iN))/G_i}$$

which is a quotient of the abelian group G_{i+1}/G_i, so is abelian. Therefore G/N is soluble.

(3) There exist two series

$$1 = N_0 \lhd N_1 \lhd \cdots \lhd N_r = N$$

$$N/N = G_0/N \lhd G_1/N \lhd \cdots \lhd G_s/N = G/N$$

with abelian quotients. Consider the series of G given by

$$1 = N_0 \lhd N_1 \lhd \cdots \lhd N_r = N = G_0 \lhd G_1 \lhd \cdots \lhd G_s = G.$$

The quotients are either N_{i+1}/N_i (which is abelian) or G_{i+1}/G_i, which is isomorphic to

$$\frac{G_{i+1}/N}{G_i/N}$$

and again is abelian. Therefore G is soluble.

Let us say that a group G is an *extension* of a group A by a group B if G has a normal subgroup N isomorphic to A such that G/N is isomorphic to B. Then we may sum up the three properties of the above theorem by saying that *the class of soluble groups is closed under taking subgroups, quotients, and extensions.* The class of abelian groups is closed under subgroups and quotients, but not extensions; and it is largely for this reason that we are led to define soluble groups.

Simple groups

We turn to groups which are in a sense the opposite of soluble.

Definition. A group G is *simple* if its only normal subgroups are 1 and G.

Every cyclic group of prime order is simple, since it has no subgroups other than 1 and G (and hence no other *normal* subgroups). These groups are also abelian, hence soluble. They are in fact the only simple soluble groups:

13.3 Theorem. *A soluble group is simple if and only if it is cyclic of prime order.*

Proof. If G is a simple soluble group it has a series

$$1 = G_0 \lhd G_1 \lhd \cdots \lhd G_n = G$$

where by deleting repeats we may assume $G_{i+1} \neq G_i$. Then G_{n-1} is a proper normal subgroup of G, which is simple, so $G_{n-1} = 1$ and $G = G_n/G_{n-1}$ which is abelian. Since every subgroup of an abelian group is normal, and every element of G generates a cyclic subgroup, G must be cyclic with no non-trivial proper subgroups. Hence G has prime order.

The converse implication is trivial.

Simple groups play an important role in finite group theory. They are in a sense the fundamental units from which all finite groups are made. Specifically, every finite group has a series of subgroups like (1) whose quotients are simple; and these simple groups depend only on the group and not on the series chosen (this is the content of the *Jordan-Hölder theorem*).

We shall not need to know much about simple groups, intriguing as they are. We require just one result:

13.4 Theorem. *If* $n \geq 5$ *then the alternating group* $\mathbf{A}_n$ *of degree n is simple.*

Proof. Suppose that $1 \neq N \lhd \mathbf{A}_n$. Our strategy will be as follows: first, observe that if N contains a 3-cycle then it contains *all* 3-cycles, and since the 3-cycles generate $\mathbf{A}_n$, we must have $N = \mathbf{A}_n$. Second, prove that N *must* contain a 3-cycle. (It is here that we need $n \geq 5$.)

Suppose, then, that N contains a 3-cycle; without loss of generality N contains (123). Now for any $k > 3$ the cycle $(32k)$ is an even permutation, so lies in $\mathbf{A}_n$, and therefore

$$(32k)^{-1}(123)(32k) = (1k2)$$

lies in N. Hence N contains $(1k2)^2 = (12k)$ for all $k \geq 3$. Now the symmetric group is generated by all 2-cycles of the form $(1i)$ for $i = 2, \cdots, n$. Since $\mathbf{A}_n$ is the set of products of an *even* number of these, it is generated by all elements of the form

$$(1i)(1j) = (1ij).$$

But for $i \neq 2$ we have

$$(1ij) = (12j)(12i)(12j),$$

so that $\mathbf{A}_n$ is generated by all the cycles $(12k)$, which shows that $N = \mathbf{A}_n$.

It remains to show that N must contain at least one 3-cycle. We do this by an analysis into cases.

1 Suppose N contains an element

$$x = abc\cdots$$

where $a, b, c, \cdots$ are disjoint cycles and

$$a = (a_1\cdots a_m) \quad (m \geq 4).$$

Let $t = (a_1 a_2 a_3)$. Then N contains $t^{-1}xt$. Since t commutes with $b, c, \cdots$ (disjointness of cycles) it follows that

$$t^{-1}xt = (t^{-1}at)bc\cdots = z \quad \text{(say)}$$

so that N contains

$$zx^{-1} = (a_1 a_3 a_m)$$

which is a 3-cycle.

2 Now suppose N contains an element involving at least two 3-cycles. Without loss of generality N contains

$$x = (123)(456)y$$

where y is a permutation fixing 1, 2, 3, 4, 5, 6. Let $t = (234)$. Then N contains

$$(t^{-1}xt)x^{-1} = (12436).$$

Then by case (1) N contains a 3-cycle.

3 If N contains no elements covered by the previous cases, then every element of N involves either just one 3-cycle or is a product of disjoint 2-cycles. The first possibility can be dealt with by considering

$$x = (123)p$$

where p commutes with x and $p^2 = 1$. Then N contains

$$x^2 = (132)p^2 = (132)$$

which is a 3-cycle.

4 There remains the case when every element of N is a product of disjoint 2-cycles. (This actually occurs when $n = 4$, giving the four-group **V**.) But as $n \geq 5$ we can assume that N contains

$$x = (12)(34)p$$

where p fixes 1, 2, 3, 4. If we let $t = (234)$ then N contains

$$(t^{-1}xt)x^{-1} = (14)(23)$$

and if $u = (145)$ N contains

$$(u^{-1}xu)x^{-1} = (45)(23)$$

so that N contains

$$(45)(23)(14)(23) = (145)$$

contradicting the assumption that every element of N is a product of disjoint 2-cycles.

Hence $\mathbf{A}_n$ is simple if $n \geq 5$.

In fact $\mathbf{A}_5$ is the smallest simple non-abelian group, which was first proved by Galois.

From this theorem we deduce:

13.5 Corollary. *The symmetric group $\mathbf{S}_n$ of degree n is not soluble if $n \geq 5$.*

Proof. If $\mathbf{S}_n$ were soluble then $\mathbf{A}_n$ would be soluble by 13.2 and simple by 13.4, hence of prime order by 13.3. But $|\mathbf{A}_n| = \frac{1}{2}(n!)$ is not prime if $n \geq 5$.

While on the topic of the symmetric group, we prove a simple result which we shall need in Chapter 14.

13.6 Lemma. *For any n the symmetric group $\mathbf{S}_n$ is generated by the cycles $(12\cdots n)$ and (12).*

Proof. Let $c = (12\cdots n)$, $t = (12)$, and let G be the group generated by c and t. Then G contains $c^{-1}tc = (23)$, hence $c^{-1}(23)c = (34)$, $\cdots$ and hence all transpositions $(m, m+1)$. Then G contains $(12)(23)(12) = (13)$, $(13)(34)(13) = (14)$, $\cdots$ and therefore contains all transpositions $(1m)$. But then G contains all $(1m)(1r)(1m) = (mr)$. But every element of S_n is a product of transpositions, so that $G = \mathbf{S}_n$.

p-groups

We begin by recalling several ideas from group theory.

Definition. Elements a and b of a group G are *conjugate* in G if there exists $g \in G$ such that $a = g^{-1}bg$.

Conjugacy is an equivalence relation; the equivalence classes are the *conjugacy classes* of G.

If the conjugacy classes of G are $C_1, \cdots, C_r$ then one of them, say C_1, contains only the identity element of G. Therefore $|C_1| = 1$. Since the conjugacy classes form a partition of G we have

$$|G| = 1 + |C_2| + \cdots + |C_r| \qquad (2)$$

which is the *class equation* for G.

Definition. If G is a group and $x \in G$ then the *centralizer* $C_G(x)$ of x in G is the set of all $g \in G$ for which

$$xg = gx.$$

It is always a subgroup of G.

There is a useful connection between centralizers and conjugacy classes.

13.7 Lemma. *If G is a group and $x \in G$ then the number of elements in the conjugacy class of x is the index of $C_G(x)$ in G.*

Proof. The equation

$$g^{-1}xg = h^{-1}xh$$

holds if and only if

$$hg^{-1}x = xhg^{-1},$$

which means that

$$hg^{-1} \in C_G(x),$$

i.e. that h and g lie in the same coset of $C_G(x)$ in G. The number of these cosets is the index of $C_G(x)$ in G, so the lemma is proved.

13.8 Corollary. *The number of elements in any conjugacy class of a finite group G divides the order of G.*

We now introduce the class of p-groups.

Definition. Let p be a prime. A finite group G is a *p-group* if its order is a power of p.

For example, the dihedral group $\mathbf{D}_8$ is a 2-group. If $n \geq 3$ the symmetric group $\mathbf{S}_n$ is never a p-group for any prime p.

In order to state an important property of p-groups we need another definition.

Definition. The *centre* $Z(G)$ of a group G is the set of elements $x \in G$ such that $xg = gx$ for all $g \in G$.

The centre of G is a normal subgroup of G. Many groups have trivial centre, for example $Z(\mathbf{S}_3) = 1$; whereas abelian groups go to the other extreme and have $Z(G) = G$. The existence of a non-trivial centre is often useful in group theory, and p-groups are well-behaved in this respect.

13.9 Theorem. *If $G \neq 1$ is a finite p-group then G has non-trivial centre.*

Proof. The class equation (2) of G reads

$$p^n = |G| = 1 + |C_2| + \cdots + |C_r|$$

and by 13.8 we have

$$|C_i| = p^{n_i}$$

for some $n_i \geq 0$. Now p divides the right-hand side, so that at least $p - 1$ values n_i must be equal to 1. But if x lies in a conjugacy class with only 1 element, then

$$g^{-1}xg = x$$

for all $g \in G$, i.e. $gx = xg$. Hence $x \in Z(G)$. Therefore $Z(G) \neq 1$.

From this we easily deduce:

13.10 Lemma. *If G is a finite p-group of order p^n then G has* *a series of normal subgroups*

$$1 = G_0 \subseteq G_1 \subseteq \cdots \subseteq G_n = G$$

such that $|G_i| = p^i$ *for all* $i = 0, \cdots, n$.

Proof. We use induction on n. If $n = 0$ all is clear. If not, let $Z = Z(G) \neq 1$ by 13.9. Since Z is an abelian group of order p^m it has an element of order p. The cyclic subgroup K generated by such an element has order p and is *normal* in G since $K \subseteq Z$. Now G/K is a p-group of order p^{n-1}, and by induction there is a series of normal subgroups

$$K/K = G_1/K \subseteq \cdots \subseteq G_n/K$$

where $|G_i/K| = p^{i-1}$. But then $|G_i| = p^i$ and $G_i \lhd G$. If we let $G_0 = 1$ the result follows.

13.11 Corollary. *Every finite p-group is soluble.*

Proof. The quotients G_{i+1}/G_i of the series afforded by 13.10 are of order p, hence cyclic and abelian.

In 1872 the Norwegian mathematician L. Sylow discovered some very fundamental theorems about the existence of p-groups inside given finite groups. We shall need one of his results in the chapter on solution of equations by radicals, and again in proving the 'fundamental theorem of algebra'. We state all of his results, though we shall prove only the one we require.

13.12 Theorem (*Sylow*). *Let G be a finite group of order $p^\alpha r$ where p is prime and does not divide r. Then*
    (1) *G possesses at least one subgroup of order p^α,*
    (2) *All such subgroups are conjugate in G,*
    (3) *Any p-subgroup of G is contained in one of order p^α,*
    (4) *The number of subgroups of G of order p^α leaves remainder 1 on division by p.*

This result motivates the:

Definition. If G is a finite group of order $p^\alpha r$ where p is prime and does not divide r, then a *Sylow p-subgroup* of G is a subgroup of G of order p^α.

In this terminology Theorem 13.12 says that for finite groups Sylow p-subgroups exist for all primes p, are all conjugate, are the maximal p-subgroups of G, and occur in numbers restricted by condition (4).

We want to prove part (1) of this theorem; and we need an auxiliary result to do so.

13.13 Lemma. *If A is a finite abelian group whose order is divisible by a prime p then A has an element of order p.*

Proof. We use induction on $|A|$. If $|A|$ is prime the result follows. Otherwise take a proper subgroup M of A whose order m is maximal. If p divides m we are home by induction, so we may assume that p does not divide m. Let t be in A but not in M, and let T be the cyclic subgroup generated by T. Then MT is a subgroup of A, larger than M, so by maximality $A = MT$. From the first isomorphism theorem

$$|MT| = |M||T|/|M \cap T|$$

so that p divides the order r of T. Since T is cyclic the element $t^{r/p}$ has order p. The lemma follows.

Proof of 13.12 part 1. We use induction on $|G|$. The theorem is obviously true for $|G| = 1$ or 2. Let $C_1, \cdots, C_s$ be the conjugacy classes of G, and let $c_i = |C_i|$. The class equation of G is

$$p^\alpha r = c_1 + \cdots + c_s. \tag{3}$$

Let Z_i denote the centralizer in G of some element $x_i \in C_i$, and let $n_i = |Z_i|$. By 13.7 we have

$$n_i = p^\alpha r/c_i. \tag{4}$$

Suppose first that some c_i is greater than 1 and not divisible by p. Then by (4) $n_i < p^\alpha r$ and is divisible by p^α. Hence by induction Z_i contains a subgroup of order p^α. Therefore we may assume that for all $i = 1, \cdots, s$ either $c_i = 1$ or p divides c_i. Let $z = |Z(G)|$. As in 13.9, z is the number of values of i such that $c_i = 1$. So

$$p^\alpha r = z + kp$$

for some integer p. Hence p divides z, and G has a non-trivial centre Z such that p divides $|Z|$. By 13.13 Z has an element of order p, which generates a subgroup P of G of order p. Since $P \subseteq Z$ it follows that $P \lhd G$. By induction G/P contains a subgroup S/P of order $p^{\alpha-1}$, whence S is a subgroup of G of order p^α and the theorem is proved.

From this result we can derive a theorem of Cauchy (which historically preceded Sylow's theorem):

13.14 Theorem (*Cauchy*). *If a prime p divides the order of a finite group G then G has an element of order p.*

Proof. Let S be a Sylow p-subgroup of G, so that $S \neq 1$. By 13.10 S has a normal subgroup of order p. Any non-identity element of this has order p.

Example. Let $G = \mathbf{S}_4$, so that $|G| = 24$. According to Sylow's theorem G must have subgroups of orders 3 and 8. Subgroups of order 3 are easy to find: any 3-cycle, such as (123) or (134) or (234), generates such a group. We shall find a subgroup of order 8. Let $\mathbf{V}$ be the four-group, which is normal in G. Let t be any 2-cycle, generating a subgroup T of order 2. Then $\mathbf{V} \cap T = 1$, and $\mathbf{V}T$ is a subgroup of order 8.

Analogues of Sylow's theorem do not work as soon as we go beyond prime powers. Exercise 13.8 illustrates this point.

Exercises

13.1   Show that the general dihedral group

$$\mathbf{D}_{2n} = \langle a, b : a^n = b^2 = 1, b^{-1}ab = a^{-1} \rangle$$

is a soluble group.

13.2 Prove that $\mathbf{S}_n$ is not soluble for $n \geq 5$ using only the simplicity of $\mathbf{A}_5$.

13.3 Prove that a normal subgroup of a group is a union of conjugacy classes. Find the conjugacy classes of $\mathbf{A}_5$ (using the cycle type of the permutations) and hence show that $\mathbf{A}_5$ is simple.

13.4 Prove that $\mathbf{S}_n$ is generated by the 2-cycles $(12), \cdots, (1n)$.

13.5 Find Sylow 2-, 3-, and 5-subgroups of $\mathbf{S}_5$.

13.6 Prove that a finite group G is a p-group if and only if every element has p-power order.

13.7 If the point (α, β) is constructible by ruler and compasses from $(0, 0)$ and $(1, 0)$ show that the Galois groups of $\mathbf{Q}(\alpha): \mathbf{Q}$ and $\mathbf{Q}(\beta): \mathbf{Q}$ are 2-groups.

13.8 Show that $\mathbf{A}_5$ has no subgroup of order 15.

13.9 Show that a subgroup or a quotient of a p-group is again a p-group. Show that an extension of a p-group by a p-group is a p-group.

13.10 Show that $\mathbf{S}_n$ has trivial centre if $n \geq 3$.

13.11 Prove that every group of order p^2 (p prime) is abelian. Hence show that there are exactly two non-isomorphic groups of order p^2 for any prime p.

13.12 Find the conjugacy classes of the dihedral group $\mathbf{D}_{2n}$. Work out the centralizers of selected elements, one from each conjugacy class, and check Lemma 13.7.

13.13 If $x, g \in G$ show that $C_G(g^{-1}xg) = g^{-1}C_G(x)g$.

13.14 Mark the following true or false.
  (a) Every soluble group is a p-group.
  (b) Every Sylow subgroup of a finite group is soluble.
  (c) A direct product of soluble groups is soluble.
  (d) A simple soluble group is cyclic.
  (e) Every cyclic group is simple.
  (f) The symmetric group $\mathbf{S}_n$ is simple if $n \geq 5$.
  (g) The conjugacy classes of a group G are all sub-groups of G.
  (h) Every finite group possesses Sylow p-subgroups for any prime p.
  (i) Every finite non-trivial p-group has non-trivial centre.
  (j) Every simple p-group is abelian.

CHAPTER FOURTEEN

Solution of equations by radicals

The historical aspects of the problem of solving polynomial equations by radicals have been discussed in the introduction. The object of this chapter is to use the Galois correspondence to give a condition which must be satisfied by an equation soluble by radicals, namely: the associated Galois group must be a soluble group. We then construct a quintic polynomial whose Galois group is not soluble, which shows that the quintic equation cannot be solved by radicals.

Solubility of the Galois group is also a *sufficient* condition for an equation to be soluble by radicals. We defer consideration of this result to the next chapter.

Radical extensions

Some care is needed in formalizing the idea of 'solubility by radicals'. We begin from the point of view of field extensions.

Definition. An extension $L:K$ is *radical* if $L = K(\alpha_1, \cdots, \alpha_m)$ where for each $i = 1, \cdots, m$ there exists an integer $n(i)$ such that

$$\alpha_i^{n(i)} \in K(\alpha_1, \cdots, \alpha_{i-1}).$$

The elements α_i are said to form a *radical sequence* for $L:K$.

Informally, a radical extension is obtained by a sequence of adjunctions of nth roots, for various n. For example the radical expression

$$\sqrt[3]{11}\,\sqrt[5]{\frac{7+\sqrt{3}}{2}}+\sqrt[4]{1+\sqrt[3]{4}}$$

is contained in a radical extension $\mathbf{Q}(\alpha, \beta, \gamma, \delta, \varepsilon)$ of $\mathbf{Q}$, where $\alpha^3 = 11$, $\beta^2 = 3$, $\gamma^5 = (7+\beta)/2$, $\delta^3 = 4$, $\varepsilon^4 = 1+\delta$.

It is clear that any radical expression, in the sense of the Introduction, is contained in some radical extension.

A polynomial should be soluble by radicals provided all of its zeros are radical expressions over the ground field.

Definition. Let f be a polynomial over a field K of characteristic zero, and let Σ be a splitting field for f over K. We say that f is *soluble by radicals* if there exists a field M containing Σ such that $M:K$ is a radical extension.

There are several points to be made about this definition. The first is the restriction to characteristic zero: for technical reasons it is customary to use a slightly wider definition of solubility by radicals when fields of characteristic $p > 0$ are under consideration. To keep our treatment simple we shall assume characteristic zero throughout this chapter, while occasionally indicating how to modify the treatment for characteristic p.

The second is to emphasize that $\Sigma:K$ need not itself be radical. We want everything in the splitting field to be expressible by radicals, but it is pointless to expect everything expressible by the same radicals to be inside the splitting field. If $M:K$ is radical and L is an intermediate field, then $L:K$ need not be radical.

The third is that we require *all* zeros of f to be expressible by radicals. It is possible for some zeros to be expressible by radicals, whilst others are not – simply take a product of two polynomials, one soluble by radicals and one not. However, if an *irreducible* polynomial f has one zero expressible by radicals then *all* the zeros must be, by a simple argument based on Theorem 3.8.

The main theorem we shall prove is:

14.1 Theorem. *If K is a field of characteristic zero and $K \subseteq L \subseteq M$ where $M:K$ is a radical extension, then the Galois group of $L:K$ is a soluble group.*

The curious word 'soluble' for groups arises in this context: a soluble (by radicals) polynomial has a soluble Galois group (of a splitting field over the base field).

The proof of this result is far from straightforward, and we must spend some time on preliminaries.

14.2 Lemma. *Suppose that $L:K$ is finite and M is a normal closure of $L:K$. Then M is generated by subfields $L_1, \cdots, L_s$ containing K, such that each extension $L_i:K$ is isomorphic to $L:K$.*

Proof. By 4.4, $L = K(\alpha_1, \cdots, \alpha_r)$ for algebraic elements $\alpha_1, \cdots, \alpha_r$ over K. Let m_i be the minimum polynomial of α_i over K, and let N be a splitting field for $f = m_1 \cdots m_r$ over K, containing L. Then N is a normal closure of $L:K$ by the proof of 10.3. Also by 10.3 the extensions $M:K$ and $N:K$ are isomorphic, so we may assume that $M = N$. Now if β_i is any zero of m_i it follows from 3.9 that the extensions

$$K(\alpha_1, \cdots, \alpha_i, \cdots, \alpha_r):K$$

$$K(\alpha_1, \cdots, \beta_i, \cdots, \alpha_r):K$$

are isomorphic. But as β_i and i vary, the latter extensions generate M. The lemma follows.

We need this result to prove:

14.3 Lemma. *If $L:K$ is a radical extension and M is a normal closure of $L:K$ then $M:K$ is radical.*

Proof. M is generated by $L_1, \cdots, L_s$ as in 14.2, and the extensions $L_i:K$ are all isomorphic to $L:K$, so are radical. By induction it suffices to prove that if R is generated by R_1 and R_2 where $R_1:K$ and $R_2:K$ are radical, then $R:K$ is radical. Let $R_1 = K(\alpha_1, \cdots, \alpha_m)$ and $R_2 = K(\beta_1, \cdots, \beta_n)$ where the α_i and β_j are radical sequences. Then the combined sequence

$$\alpha_1, \cdots, \alpha_m, \beta_1, \cdots, \beta_n$$

is a radical sequence, so that

$$K(\alpha_1, \cdots, \alpha_m, \beta_1, \cdots, \beta_n):K$$

is radical. But this is $R:K$.

The next two lemmas show that certain Galois groups are abelian.

14.4 Lemma. *Let K be a field of characteristic zero, and let L be a splitting field for $t^p - 1$ over K, where p is prime. Then the Galois group of $L:K$ is abelian.*

Proof. The derivative of $t^p - 1$ is pt^{p-1} so the polynomial has no multiple zeros in L. Clearly its zeros form a group under multiplication; this has order p since the zeros are distinct, so is cyclic. Let ε be a generator of this group. Then $L = K(\varepsilon)$ so that any K-automorphism of L is determined by its effect on ε. Further, K-automorphisms permute the zeros of $t^p - 1$. Hence any K-automorphism of L is of the form

$$\alpha_j: \varepsilon \to \varepsilon^j.$$

But then $\alpha_i \alpha_j$ and $\alpha_j \alpha_i$ both send ε to ε^{ij}, so the Galois group is abelian.

14.5 Lemma. *Let K be a field of characteristic zero in which $t^n - 1$ splits. Let $a \in K$, and let L be a splitting field for $t^n - a$ over K. Then the Galois group of $L:K$ is abelian.*

Proof. Let α be any zero of $t^n - a$. Since $t^n - 1$ splits in K, the general zero of $t^n - a$ is $\varepsilon\alpha$ where ε is a zero of $t^n - 1$ in K. Since $L = K(\alpha)$ any K-automorphism of L is determined by its effect on α. Given two K-automorphisms

$$\phi : \alpha \to \varepsilon\alpha$$

$$\psi : \alpha \to \eta\alpha$$

where ε and $\eta \in K$, then $\phi\psi(\alpha) = \varepsilon\eta\alpha = \eta\varepsilon\alpha = \psi\phi(\alpha)$. As before the Galois group is abelian.

We are now in a position to give a proof of theorem 14.1. The proof is by induction, after a series of simplifications have been effected.

Proof of 14.1

1 Let K_0 be the fixed field of the Galois group $\Gamma(L:K)$. Suppose for the moment that we could prove 14.1 with K replaced by K_0. Since the Galois group $\Gamma(L:K_0)$ is the same as $\Gamma(L:K)$, by definition, and since M is clearly radical over $K_0 \supseteq K$, we could deduce 14.1. We may therefore replace K by K_0. But $L:K_0$ is normal by 10.10, since K_0 is the fixed field of $\Sigma(L:K) = \Gamma(L:K_0)$. The point of all this is that *we may assume that $L:K$ is normal* (by rechristening K_0).

2 If N is a normal closure of $M:K$, then by 14.3 $N:K$ is radical. Replacing M by N *we may assume $M:K$ is normal.*

3 The Galois group $\Gamma(L:K)$ is a quotient group of $\Gamma(M:K)$ by 11.1 part (5). If we prove $\Gamma(M:K)$ soluble it follows that $\Gamma(L:K)$ is soluble, by 13.2 part (2). Thus *we may assume that $M = L$*; the general result will follow from this special case.

4 We have now reduced the proof to considering the following situation: $L:K$ is a normal radical extension, which is separable since the characteristic is zero. We wish to show that $\Gamma(L:K)$ is soluble. Suppose that

$$L = K(\alpha_1, \cdots, \alpha_n) \tag{1}$$

where

$$\alpha_i^{n(i)} \in K(\alpha_1, \cdots, \alpha_{i-1}). \tag{2}$$

By inserting extra α_i if necessary *we may assume that $n(i)$ is prime for all i*, while (1) and (2) still hold. In particular there is a prime p such that $\alpha_1^p \in K$.

5 We now use induction on the integer n occurring in (1). Let M_0 be a splitting field of $t^p - 1$ over L, and let M_1 be the subfield of M_0 generated by K and the zeros of $t^p - 1$ in M_0.

Since $\Gamma(M:K)$ is a quotient of $\Gamma(M_0:K)$ it suffices to prove that $\Gamma(M_0:K)$ is soluble. Now M_1 is a normal extension of K and $\Gamma(M_1:K)$ is abelian by 14.4. But

$$\Gamma(M_0:M_1) \lhd \Gamma(M_0:K)$$

by 11.1 part (4), and the Galois group $\Gamma(M_1:K)$ is isomorphic to the quotient $\Gamma(M_0:K)/(M_0:M_1)$. If we can prove that $\Gamma(M_0:M_1)$ is soluble then by 13.2 part (3) it will follow that $\Gamma(M_0:K)$ is soluble. Let $G = \Gamma(M_0:M_1)$. Now

$$M_0 = M_1(\alpha_1, \cdots, \alpha_n).$$

Let $H = (M_1(\alpha_1))^*$, the subgroup of G corresponding to $M_1(\alpha_1)$ under the Galois correspondence. Now $t^p - 1$ splits in M_1, so that $M_1(\alpha_1)$ is a splitting field for $t^p - \alpha_1^p$ over M_1. Hence it is a normal extension of M_1 (by 8.4) and has abelian Galois group (by 14.5). Hence $H \lhd G$ by 11.1 part (4). The following groups and fields now correspond under the Galois correspondence:

$$
\begin{array}{ccc}
M_0 & \leftrightarrow & G \\
| & & | \\
M_1(\alpha_1) & \leftrightarrow & H \\
| & & | \\
M_1 & \leftrightarrow & 1
\end{array}
$$

By 11.1 part (5) G/H is the Galois group $\Gamma(M_0:M_1(\alpha_1))$. But

$$M_0 = M_1(\alpha_1)(\alpha_2, \cdots, \alpha_n).$$

Since $M_0 : M_1(\alpha_1)$ is normal, it follows by induction that G/H is soluble. But H is abelian, so by 13.2 part (3) G is soluble. This completes the proof.

The *idea* of this proof is simple: a radical extension is a series of extensions by nth roots; such extensions have abelian Galois groups; so the Galois group of a radical extension is made up by fitting together a sequence of abelian groups. Unfortunately there are technical problems in carrying the proof out; we need to throw in roots of unity, and we have to make various extensions normal before the Galois correspondence can be used.

Now we translate back from fields to polynomials, and in doing so revert to Galois's original viewpoint.

Definition. Let f be a polynomial over a field K, with a splitting field Σ over K. The *Galois group of f over K* is the Galois group $\Gamma(\Sigma : K)$.

Let G be the Galois group of a polynomial f over the field K. If $\alpha \in \Sigma$ is a zero of f then $f(\alpha) = 0$, so for any $g \in G$ we have $f(g(\alpha)) = g(f(\alpha)) = 0$. Hence each element $g \in G$ induces a permutation g' of the set of zeros of f in Σ. Distinct elements of G induce distinct permutations, since Σ is generated by the zeros of f. It follows easily that the map $g \to g'$ is a group monomorphism of G into the group of all permutations of the zeros of f. In other words, *we can think of G as a group of permutations on the zeros of f.* This is how Galois *defined* the Galois group; and for many years afterwards the only groups considered by mathematicians were permutation groups. Cayley was the first to define an abstract group, although it seems that the earliest satisfactory axiom system for groups was given by Kronecker in 1870 (see Huntingdon [40]).

We may restate 14.1 as:

14.6 Theorem. *Let f be a polynomial over a field K of characteristic zero. If f is soluble by radicals then the Galois group of f over K is a soluble group.*

Thus to find a polynomial not soluble by radicals it suffices to find one whose Galois group is not soluble. There are two main ways of doing this. One is to look at the so-called 'general polynomial of degree n' (which we shall discuss in the next chapter), but this has the disadvantage that it does not show that there are polynomials with *rational* coefficients which are insoluble by radicals. The alternative approach, which we shall adopt here, is to exhibit a specific polynomial with rational coefficients whose Galois group is not soluble. Since Galois groups are hard to calculate a little low cunning is necessary, together with knowledge of the symmetric group. We must also have recourse to the 'fundamental theorem of algebra'.

An insoluble quintic

Watch carefully; there is nothing up my sleeve. . . .

14.7 Lemma. *Let p be a prime, and f an irreducible polynomial of degree p over $\mathbf{Q}$. Suppose that f has precisely 2 non-real zeros in $\mathbf{C}$. Then the Galois group of f over $\mathbf{Q}$ is the symmetric group $\mathbf{S}_p$.*

Proof. By the 'fundamental theorem of algebra' $\mathbf{C}$ contains a splitting field Σ for f. Let G be the Galois group of f over $\mathbf{Q}$, considered as a permutation group on the zeros of f. These are distinct since the characteristic is zero, so that G is a subgroup of $\mathbf{S}_p$. When we construct a splitting field for f we first adjoin an element of degree p, so that $[\Sigma:\mathbf{Q}]$ is divisible by p. By 11.1 part (1) p divides the order of G. By 13.14 G has an element of order p. But the only elements of $\mathbf{S}_p$ having order p are the p-cycles.

Complex conjugation is a $\mathbf{Q}$-automorphism of $\mathbf{C}$, and therefore induces a $\mathbf{Q}$-automorphism of Σ. This leaves the $p-2$ real zeros of f fixed, while transposing the 2 non-real zeros. Therefore G contains a 2-cycle.

By choice of notation, and if necessary taking a power of the p-cycle, we may assume that G contains the 2-cycle (12) and the p-cycle $(12\cdots p)$. But by 13.6 these generate the whole of $\mathbf{S}_p$. Therefore $G = \mathbf{S}_p$ and the lemma is proved.

We can now exhibit a quintic polynomial which is not soluble by radicals.

14.8 Theorem. *The polynomial* $t^5 - 6t + 3$ *over* $\mathbf{Q}$ *is not soluble by radicals.*

Proof. Let $f(t) = t^5 - 6t + 3$. By Eisenstein's criterion f is irreducible over $\mathbf{Q}$. We shall show that f has precisely 3 real zeros, each with multiplicity 1, and hence has 2 non-real zeros. Since 5 is prime, by 14.7 the Galois group of f over $\mathbf{Q}$ is $\mathbf{S}_5$. By 13.5 $\mathbf{S}_5$ is not soluble. By 14.6 f is not soluble by radicals.

It remains to show that f has exactly 3 real zeros, each of multiplicity 1. Now $f(-2) = -17$, $f(-1) = 8$, $f(0) = 3$, $f(1) = -2$, and $f(2) = 23$. A rough sketch of the graph of $y = f(x)$ looks like Fig. 8.

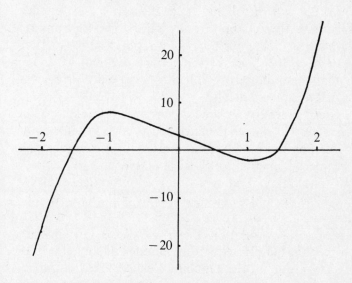

Fig. 8

This certainly appears to give only 3 real zeros, but we must be rigorous. By Rolle's theorem the zeros of f are separated by zeros of Df. And $Df = 5t^4 - 6$, which has 2 zeros at $\pm \sqrt[4]{(6/5)}$. Now f and Df are coprime so f has no repeated zeros (this also follows by irreducibility and characteristic zero) so f has at most 3 real zeros. But certainly f has at least 3 real zeros, since a continuous function defined on the real line cannot change sign except by passing through 0. Therefore f has precisely 3 real zeros and the result follows.

Of course this is not the end of the story. There are more ways of killing a quintic than choking it with radicals. Having established the inadequacy of radicals for solving the problem, it is natural to look further afield.

On a mundane level, numerical methods can be used to find the zeros (real or complex) to any required degree of accuracy. This is a useful practical method – indeed the only *practical* method. The mathematical theory of such numerical methods can be far from mundane – but from the algebraic point of view it is unilluminating.

Another way of solving the problem is to say, in effect, "What's so special about radicals?" Suppose for any real number a we define the *ultraradical* $\sqrt[*]{a}$ to be the real zero of $t^5 + t - a$. It was shown by Jerrard (see Kollros [43] p. 19) that the quintic equation can be solved by the use of radicals *and* ultraradicals.

Instead of inventing new tools we can refashion existing ones. Hermite made the remarkable discovery that the quintic equation can be solved in terms of *elliptic modular functions*, special functions of classical mathematics which arose in a quite different context (integration of algebraic functions). The method is analogous to the well-known trigonometric solution of the cubic equation. In a triumph of mathematical unification Klein succeeded in connecting together the quintic equation, elliptic functions, and the rotation group of the regular icosahedron. The latter is isomorphic to the alternating group $\mathbf{A}_5$ which we have seen plays a key part in the theory of the quintic. Klein's work helped to explain the unexpected appearance of elliptic functions in the theory of polynomial equations; these ideas were subsequently generalized by Poincaré to cover polynomials of arbitrary degree. The role of the icosahedron is explained in Klein [41].

Exercises

14.1 If you have never seen it done, or have forgotten the details, try to solve cubics and quartics by radicals. (A method is given in Chapter 16.)

14.2 Find radical extensions of $\mathbf{Q}$ containing the following elements of $\mathbf{C}$:
(a) $(\sqrt{11} - \sqrt[7]{23})/\sqrt[4]{5}$.
(b) $(\sqrt{6} + 2.\sqrt[3]{5})^4$.
(c) $(2.\sqrt[5]{5} - 4)/(\sqrt{(1 + \sqrt{99})})$.

14.3 What is the Galois group of $t^p - 1$ over $\mathbf{Q}$ for prime p?

14.4 Show that the polynomials
$$t^5 - 4t + 2$$
$$t^5 - 4t^2 + 2$$
$$t^5 - 6t^2 + 3$$
$$t^7 - 10t^5 + 15t + 5$$

over $\mathbf{Q}$ are not soluble by radicals.

14.5 Solve the sextic equation
$$t^6 + 2t^5 - 5t^4 + 9t^3 - 5t^2 + 2t + 1 = 0$$

by radicals. (Hint: put $u = t + \dfrac{1}{t}$.)

14.6 A permutation group G on a set X is *transitive* if for any $x, y \in X$ there is an element $g \in G$ such that $g(x) = y$. Show that the Galois group of an irreducible polynomial is transitive as a permutation group on the zeros of the polynomial.

14.7 If $L:K$ is a radical extension and M is an intermediate field, show that $M:K$ need not be radical.

14.8 If p is an irreducible polynomial over K with a zero expressible by radicals, prove that every zero of p is expressible by radicals.

14.9 Mark the following true or false.
  (a) Every quartic equation over a field of characteristic zero can be solved by radicals.
  (b) Every radical extension is finite.
  (c) Every finite extension is radical.
  (d) The order of the Galois group of a polynomial of degree n divides $n!$.
  (e) Any reducible quintic polynomial can be solved by radicals.
  (f) There exist quartics with Galois group $\mathbf{S}_4$.
  (g) An irreducible polynomial of degree 11 with exactly two non-real zeros has Galois group $\mathbf{S}_{11}$.
  (h) The normal closure of a radical extension is radical.
  (i) $\mathbf{A}_5$ has 60 elements.

CHAPTER FIFTEEN

The general polynomial equation

The so-called 'general' polynomial is in fact a very *special* polynomial. It is one whose coefficients do not satisfy any algebraic relations. This property makes it simpler to work with than, say, polynomials over $\mathbf{Q}$; and in particular it is easier to calculate its Galois group. As a result, we can show that the general quintic polynomial is not soluble by radicals without assuming as much group theory as we did in Chapter 14. Effectively this implies that there is no general formula by which all quintic equations can be solved in terms of radicals. Since this does not *a priori* preclude the possibility that there might exist solutions by radicals of all quintic polynomials which cannot be subsumed under a general formula, the results of this chapter are not as strong as those of the preceding chapter.

It transpires that the Galois group of the general polynomial of degree n is the whole symmetric group $\mathbf{S}_n$. This immediately shows the insolubility of the general quintic. Our knowledge of the structure of $\mathbf{S}_2$, $\mathbf{S}_3$, and $\mathbf{S}_4$ can be used to find methods of solving the general quadratic, cubic, or quartic equation.

Transcendence degrees

Up till now we have not had to deal much with transcendental extensions; indeed the assumption of *finiteness* of the extensions has been crucial to our theory. We now need to consider a wider class of extensions, which still have a flavour of finiteness.

Definition. An extension $L:K$ is *finitely generated* if $L = K(\alpha_1, \cdots, \alpha_n)$ where n is finite.

Note that the α_i may be either algebraic or transcendental over K.

Definition. If $t_1, \cdots, t_n$ are transcendental elements over a field K, all lying inside some extension L of K, then they are *independent* if there is no nontrivial polynomial p over K (in n indeterminates) such that

$$p(t_1, \cdots, t_n) = 0$$

in L.

Thus, for example, if t is transcendental over K and u is transcendental over $K(t)$ then $K(t, u)$ is a finitely generated extension of K and t, u are independent. On the other hand t and $t+1$ are both transcendental over K but are connected by a polynomial equation, so are not independent.

The next result is concerned with the structure of a finitely generated extension.

15.1 Lemma. *If $L:K$ is finitely generated then there exists an intermediate field M such that*
   (1) $M = K(\alpha_1, \cdots, \alpha_r)$ *where the α_i are independent transcendental elements over K,*
   (2) $L:M$ *is a finite extension.*

Proof. We know that $L = K(\beta_1, \cdots, \beta_n)$. If all the β_j are algebraic over K then $L:K$ is finite by 4.4 and we may take $M = K$. Otherwise some β_i is transcendental over K. Call this α_1. If $L:K(\alpha_1)$ is not finite there exists some β_k transcendental over $K(\alpha_1)$. Call this α_2. We may continue this process until $M = K(\alpha_1, \cdots, \alpha_r)$ is such that $L:M$ is finite.

By construction the α_i are independent transcendental elements over K.

A result due to Steinitz says that the integer r which gives the number of independent transcendental elements is not dependent on the choice of M.

15.2 Lemma (*Steinitz*). *With the notation of Lemma* 15.1, *if there is another intermediate field* $N = K(\beta_1, \cdots, \beta_s)$ *such that* $\beta_1, \cdots, \beta_s$ *are independent transcendental elements over* K *and* $L:N$ *is finite, then* $r = s$.

Proof. Since $[L:M]$ is finite β_1 is algebraic over M, hence there is a polynomial equation

$$p(\beta_1\alpha, \cdots, \alpha_r) = 0.$$

Some α_i, without loss of generality α_1, actually occurs in this equation. Then α_1 is algebraic over $K(\beta_1, \alpha_2, \cdots, \alpha_r)$ and $L:K(\beta_1, \alpha_2, \cdots, \alpha_r)$ is finite. Inductively we can replace successive α_i by β_i, so that $L:K(\beta_1, \cdots, \beta_r)$ is finite. If $s > r$ then β_{r+1} must be transcendental over $K(\beta_1, \cdots, \beta_r)$, a contradiction. Therefore $s \leq r$. Similarly $r \leq s$. The lemma follows.

This result means that r is well-defined in the following:

Definition. The integer r defined in 15.1 is the *transcendence degree* of $L:K$.

For example, consider $K(t, \alpha, u):K$, where t is transcendental over K, $\alpha^2 = t$, and u is transcendental over $K(t, \alpha)$. Then $M = K(t, u)$ where t and u are independent transcendental elements over K, and $K(t, \alpha, u):M = M(\alpha):M$ is finite. The transcendence degree is 2.

An easy induction shows that an extension $K(\alpha_1, \cdots, \alpha_r):K$ by independent transcendental elements α_i is isomorphic to $K(t_1, \cdots, t_r):K$ where $K(t_1, \cdots, t_r)$ is the field of rational expressions in the indeterminates t_i.

The general polynomial

Let K be any field, and let $t_1, \cdots, t_n$ be independent trans-cendental elements over K. The symmetric group $\mathbf{S}_n$ can be made to act as a group of K-automorphisms of $K(t_1, \cdots, t_n)$ by defining

$$\sigma(t_i) = t_{\sigma(i)}$$

for all $\sigma \in \mathbf{S}_n$ and extending in the obvious way to a K-automorphism. Clearly distinct elements of $\mathbf{S}_n$ give rise to distinct K-automorphisms.

The fixed field F of $\mathbf{S}_n$ obviously contains all the sym-metric polynomials in the t_i, and in particular the elemen-tary symmetric polynomials $s_r = s_r(t_1, \cdots, t_n)$. We show that these generate F.

15.3 Lemma. *With the above notation, $F = K(s_1, \cdots, s_n)$.*

Proof. First we shall show that

$$[K(t_1, \cdots, t_n):K(s_1, \cdots, s_n)] \le n!$$

by induction on n. Consider the double extension

$$K(t_1, \cdots, t_n) \supseteq K(s_1, \cdots, s_n, t_n) \supseteq K(s_1, \cdots, s_n).$$

Now $f(t_n) = 0$, where

$$f(t) = t^n - s_1 t^{n-1} + \cdots + (-1)^n s_n$$

so that

$$[K(s_1, \cdots, s_n, t_n):K(s_1, \cdots, s_n)] \le n.$$

If we let $s'_1, \cdots, s'_{n-1}$ be the elementary symmetric poly-nomials in $t_1, \cdots, t_{n-1}$ we have

$$s_j = t_n s'_{j-1} + s'_j$$

and therefore

$$K(s_1, \cdots, s_n, t_n) = K(t_n, s'_1, \cdots, s'_{n-1}).$$

Now by induction

$$[K(t_1, \cdots, t_n):K(s, \cdots, s_n, t_n)] =$$

$$= [K(t_n)(t_1, \cdots, t_{n-1}):K(t_n)(s'_1, \cdots, s'_{n-1})] \leq (n-1)!$$

so by multiplicativity of the degree the induction step goes through.

Now $K(s_1, \cdots, s_n)$ is clearly contained in the fixed field F of $\mathbf{S}_n$. By 9.4 $[K(t_1, \cdots, t_n):F] = |\mathbf{S}_n| = n!$ so by the above we must have $F = K(s_1, \cdots, s_n)$.

15.4 Corollary. *Every symmetric polynomial in $t_1, \cdots, t_n$ over K can be written as a rational expression in $s_1, \cdots, s_n$.*

Proof. Symmetric polynomials lie inside the fixed field F.

(The reader should compare this result with 2.9.)

15.5 Lemma. *With the above notation, $s_1, \cdots, s_n$ are independent transcendental elements over K.*

Proof. $K(t_1, \cdots, t_n)$ is a finite extension of $K(s_1, \cdots, s_n)$ and hence they both have the same transcendence degree over K, namely n. Therefore the s_i are independent, for otherwise the transcendence degree of $K(s_1, \cdots, s_n):K$ would be smaller than n.

Definition. Let K be a field and let $s_1, \cdots, s_n$ be independent transcendental elements over K. The *general polynomial of degree n* 'over' K is the polynomial

$$t^n - s_1 t^{n-1} + s_2 t^{n-2} - \cdots + (-1)^n s_n$$

over the field $K(s_1, \cdots, s_n)$.

15.6 Theorem. *For any field K let g be the general polynomial of degree n 'over' K and let Σ be a splitting field for g over $K(s_1, \cdots, s_n)$. Then the zeros $t_1, \cdots, t_n$ of g in Σ are independent transcendental elements over K, and the Galois group of $\Sigma:K(s_1, \cdots, s_n)$ is the full symmetric group $\mathbf{S}_n$.*

Proof. The extension $\Sigma: K(s_1, \cdots, s_n)$ is finite by 8.4, so the transcendence degree of $\Sigma: K$ is equal to that of $K(s_1, \cdots, s_n): K$, namely n. Since $\Sigma = K(t_1, \cdots, t_n)$ it follows that the t_i are independent transcendental elements over K, since any algebraic relation between them would lower the transcendence degree. The s_i are now the elementary symmetric polynomials in $t_1, \cdots, t_n$ by Chapter 2. As above, $\mathbf{S}_n$ acts as a group of automorphisms of $\Sigma = K(t_1, \cdots, t_n)$, and by 15.3 the fixed field is $K(s_1, \cdots, s_n)$. By 10.10 $\Sigma: K(s_1, \cdots, s_n)$ is separable and normal (normality also follows from the definition of Σ as a splitting field), and by 9.4 its degree is $|\mathbf{S}_n| = n!$ Then by 11.1 part (1) the Galois group has order $n!$, and contains $\mathbf{S}_n$, so is $\mathbf{S}_n$.

From 14.6 and 13.5 we deduce:

15.7 Theorem. *If K is a field of characteristic zero and $n \geq 5$, then the general polynomial of degree n 'over' K is not soluble by radicals.*

Solving quartic equations

When the general polynomial of degree n 'over' K *can* be solved by radicals, it is easy to deduce a solution by radicals of any polynomial of degree n over K, by substituting elements of K for $s_1, \cdots, s_n$ in the solution. This is the source of the 'generality' of the general polynomial. From 15.7 the best we can hope for is a solution of polynomials of degree ≤ 4. We fulfil this hope by analysing the structure of $\mathbf{S}_n$ for $n \leq 4$, and using a converse to Theorem 14.6, which we proceed to prove.

Definition Let $L: K$ be a finite normal extension with Galois group G. The *norm* of an element $a \in L$ is

$$N(a) = \tau_1(a)\tau_2(a)\cdots\tau_n(a)$$

where $\tau_1, \cdots, \tau_n$ are the elements of G.

Clearly $N(a)$ lies in the fixed field of G (use 9.3) so if the extension is also separable then $N(a) \in K$.

The next result is traditionally referred to as *Hilbert's Theorem 90* from its appearance in his 1893 report on algebraic numbers.

15.8 Theorem. *Let $L:K$ be a finite normal extension with cyclic Galois group G generated by an element τ. Then $a \in L$ has norm $N(a) = 1$ if and only if*

$$a = b/\tau(b)$$

for some $b \in L$, where $b \neq 0$.

Proof. If $a = b/\tau(b)$ and $b \neq 0$ then if $|G| = n$ we have

$$N(a) = a . \tau(a) . \tau^2(a) \cdots \tau^{n-1}(a)$$

$$= \frac{b}{\tau(b)} . \frac{\tau(b)}{\tau^2(b)} . \frac{\tau^2(b)}{\tau^3(b)} \cdots \frac{\tau^{n-1}(b)}{\tau^n(b)}$$

$$= 1 \text{ since } \tau^n = 1.$$

Conversely, suppose that $N(a) = 1$. Let $c \in L$, and define

$$d_0 = ac$$

$$d_1 = (a . \tau(a))\tau(c)$$

$$\cdots$$

$$d_i = (a . \tau(a) \cdots \tau^i(a))\tau^i(c)$$

for $0 \leq i \leq n-1$. Then

$$d_{n-1} = N(a)\tau^{n-1}(c) = \tau^{n-1}(c).$$

Further,

$$d_{i+1} = a . \tau(d_i) \qquad 0 \leq i \leq n-2.$$

Define

$$b = d_0 + d_1 + \cdots + d_{n-1}.$$

We shall choose c in such a way that $b \neq 0$. Suppose on the contrary that $b = 0$ for all choices of c. Then for any

$c \in L$ we have

$$\lambda_0 \tau^0(c) + \lambda_1 \tau(c) + \cdots + \lambda_{n-1} \tau^{n-1}(c) = 0$$

where

$$\lambda_i = a . \tau(a) \cdots \tau^i(a)$$

belongs to L. Hence the distinct automorphisms τ^i are linearly dependent over L, contrary to 9.1.

Hence we can choose c so that $b \neq 0$. But now

$$\tau(b) = \tau(d_0) + \cdots + \tau(d_{n-1})$$

$$= \frac{1}{a}(d_1 + \cdots + d_{n-1}) + \tau^n(c)$$

$$= \frac{1}{a}(d_0 + \cdots + d_{n-1})$$

$$= b/a.$$

Therefore $a = b/\tau(b)$ as claimed.

15.9 Theorem. *Suppose that $L:K$ is a finite separable normal extension with Galois group G cyclic of prime order p, generated by τ. Assume that the characteristic of K is 0 or prime to p, and that $t^p - 1$ splits in K. Then $L = K(\alpha)$ where α is a zero of an irreducible polynomial $t^p - a$ over K, for some $a \in K$.*

Proof. The p zeros of $t^p - 1$ form a group of order p, which must be cyclic, so the zeros of $t^p - 1$ are the powers of some $\varepsilon \in K$, where $\varepsilon^p = 1$. But then

$$N(\varepsilon) = \varepsilon \cdots \varepsilon = 1$$

since $\varepsilon \in K$ so that $\tau^i(\varepsilon) = \varepsilon$ for all i. By 15.8 we have

$$\varepsilon = \alpha/\tau(\alpha)$$

for some $\alpha \in L$. It follows that

$$\tau(\alpha) = \varepsilon^{-1}\alpha, \; \tau^2(\alpha) = \varepsilon^{-2}\alpha, \; \cdots$$

and $a = \alpha^p$ is fixed by G, and so lies in K. Now $K(\alpha)$ is a splitting field for $t^p - a$ over K. The K-automorphisms $1, \tau, \cdots, \tau^{p-1}$ map α to distinct elements, so they give p distinct K-automorphisms of $K(\alpha)$. By 11.1 part (1) $[K(\alpha):K] \geq p$. But $[L:K] = |G| = p$, so $L = K(\alpha)$. Hence $t^n - a$ is the minimum polynomial of α over K (otherwise we should have $[K(\alpha):K] < p$) so is irreducible over K.

We can now prove the promised converse to 14.6.

15.10 Theorem. *Let K be a field of characteristic 0 and let $L:K$ be a finite normal extension with soluble Galois group G. Then there exists an extension R of L such that $R:K$ is radical.*

Proof. All extensions are separable since the characteristic is 0. We use induction on $|G|$. The result is clear when $|G| = 1$. If $|G| \neq 1$ we take a maximal proper normal subgroup H of G (which exists since G is a finite group). Then G/H is simple (since H is maximal) and soluble by 13.2 part (2). By 13.3 G/H is cyclic of prime order p. Let N be a splitting field over L of $t^p - 1$. Then $N:K$ is normal; for by 8.4 L is a splitting field over K of some polynomial f, and so N is a splitting field over L of $(t^p - 1).f$, so $N:K$ is normal by 8.4 again. The Galois group of $N:L$ is abelian by 14.4, and by 11.1 part (5) $\Gamma(L:K)$ is isomorphic to $\Gamma(N:K)/\Gamma(N:L)$. By 13.2 part (3) $\Gamma(N:K)$ is soluble. Let M be the subfield of N generated by K and the zeros of $t^p - 1$. Then $N:M$ is normal. Now $M:K$ is clearly radical, and since $L \subseteq N$ the desired result will follow provided we can find an extension R of N such that $R:M$ is radical.

We claim that the Galois group of $N:M$ is isomorphic to a subgroup of G. Let us map any M-automorphism τ of N into its restriction $\tau|_L$. Since $L:K$ is normal $\tau|_L$ is a K-automorphism of L, and we have a group homomorphism

$$\phi : \Gamma(N:M) \to \Gamma(L:K).$$

If $\tau \in \ker(\phi)$ then τ fixes all elements of M and L, which generate N. Therefore $\tau = 1$, so ϕ is a monomorphism, and $\Gamma(N:M)$ is isomorphic to a subgroup J of $\Gamma(L:K)$.

If $J = \phi(\Gamma(N:M))$ is a proper subgroup of G then by induction there is an extension R of N such that $R:M$ is radical.

The remaining possibility is that $J = G$. Then we can find a subgroup $I \lhd \Gamma(N:M)$ of index p, namely $I = \phi^{-1}(H)$. Let P be the fixed field $I^{\dagger}$. Then $[P:M] = p$ by 11.1 part (3), $P:M$ is normal by 11.1 part (4), and $t^p - 1$ splits in M. By 15.9 $P = M(\alpha)$ where $\alpha^p = a \in M$. But $N:P$ is a normal extension with soluble Galois group of order smaller than $|G|$, so by induction there exists an extension R of N such that $R:P$ is radical. But then $R:M$ is radical, and the theorem is proved.

Remark. To deal with field of characteristic $p > 0$ we must define radical extensions differently. As well as adjoining elements α such that α^n lies in the given field we must also allow adjunction of elements α such that $\alpha^p - \alpha$ lies in the given field (where p is the same as the characteristic). It is then true that a polynomial is soluble by radicals if and only if its Galois group is soluble. The proof differs in considering extensions of degree p over fields of characteristic p, when 15.9 breaks down; and extensions of the second type above come in. If we do not modify the definition of solubility by radicals then although every soluble polynomial has soluble group the converse need not hold – indeed some quadratic polynomials, having *abelian* Galois group, will not be soluble by radicals. (See Exercises 3.12 and 3.13.)

Since a splitting field is always a normal extension, we have:

15.11 Theorem. *Over a field of characteristic zero, a polynomial is soluble by radicals if and only if it has a soluble Galois group.*

Proof. Use 14.6 and 15.10.

The general polynomial of degree n has Galois group $\mathbf{S}_n$, and we know that for $n \leq 4$ this is soluble (see examples at the beginning of Chapter 13). Hence *for a field of characteristic zero the general polynomial of degree ≤ 4 can be solved by radicals.*

We can use our insight into the structure of the symmetric group to find out *how*.

1 *Linear*

$$t - s_1.$$

Trivially $t_1 = s_1$ is a zero.

2 *Quadratic*

$$t^2 - s_1 t + s_2.$$

Let the zeros be t_1 and t_2. The Galois group $\mathbf{S}_2$ consists of the identity and a map interchanging t_1 and t_2. Hence

$$(t_1 - t_2)^2$$

is fixed by $\mathbf{S}_2$, so lies in $K(s_1, s_2)$. By explicit calculation

$$(t_1 - t_2)^2 = s_1^2 - 4s_2.$$

Hence

$$t_1 - t_2 = \pm\sqrt{(s_1^2 - 4s_2)}$$

$$t_1 + t_2 = s_1$$

and we have the familiar formula

$$t_1, t_2 = \frac{s_1 \pm \sqrt{(s_1^2 - 4s_2)}}{2}.$$

3 *Cubic*

$$t^3 - s_1 t^2 + s_2 t^3 - s_3.$$

Let the zeros be t_1, t_2, t_3. The Galois group $\mathbf{S}_3$ has a series

$$1 \lhd \mathbf{A}_3 \lhd \mathbf{S}_3$$

with abelian quotients.

Adjoin an element $\omega \neq 1$ such that $\omega^3 = 1$. Consider

$$y = t_1 + \omega t_2 + \omega^2 t_3.$$

The elements of $\mathbf{A}_3$ permute t_1, t_2, and t_3 cyclically, and therefore multiply y by a power of ω. Hence y^3 is fixed by $\mathbf{A}_3$. Similarly if

$$z = t_1 + \omega^2 t_2 + \omega t_3$$

then z^3 is fixed by $\mathbf{A}_3$.

Now any odd permutation in $\mathbf{S}_3$ interchanges y^3 and z^3, so that

$$y^3 + z^3 \quad \text{and} \quad y^3 z^3$$

are fixed by the whole of $\mathbf{S}_3$, hence lie in $K(s_1, s_2, s_3)$. (Explicit formulae are given in the final section of this chapter.) Hence y^3 and z^3 are zeros of a quadratic over $K(s_1, s_2, s_3)$ which can be solved as in part (2). Taking cube roots we know y and z. But since

$$s_1 = t_1 + t_2 + t_3$$

it follows that

$$t_1 = \tfrac{1}{3}(s_1 + y + z)$$
$$t_2 = \tfrac{1}{3}(s_1 + \omega^2 y + \omega z)$$
$$t_3 = \tfrac{1}{3}(s_1 + \omega y + \omega^2 z).$$

4 *Quartic*

$$t^4 - s_1 t^3 + s_2 t^2 - s_3 t + s_4.$$

Let the zeros by t_1, t_2, t_3, t_4. The Galois group $\mathbf{S}_4$ has a series

$$1 \lhd \mathbf{V} \lhd \mathbf{A}_4 \lhd \mathbf{S}_4$$

with abelian quotients, where

$$\mathbf{V} = \{1, (12)(34), (13)(24), (14)(23)\}.$$

It is therefore natural to consider the three expressions

$$y_1 = (t_1 + t_2)(t_3 + t_4)$$
$$y_2 = (t_1 + t_3)(t_2 + t_4)$$
$$y_3 = (t_1 + t_4)(t_2 + t_3).$$

These are permuted amongst themselves by any permutation in $\mathbf{S}_4$, so that all the elementary symmetric polynomials in y_1, y_2, y_3 lie in $K(s_1, s_2, s_3, s_4)$. (Explicit formulae are indicated below.) Then y_1, y_2, y_3 are the zeros of a certain cubic polynomial over $K(s_1, s_2, s_3, s_4)$ called the *resolvent cubic*. Since

$$t_1 + t_2 + t_3 + t_4 = s_1$$

we can find three quadratic polynomials whose zeros are $t_1 + t_2$ and $t_3 + t_4$, $t_1 + t_3$ and $t_2 + t_4$, $t_1 + t_4$ and $t_2 + t_3$. From these it is easy to find t_1, t_2, t_3, t_4.

For completeness we shall state the explicit formulae whose existence is alluded to above. For details of the calculations, see Van der Waerden [5] pp. 177 ff.

Cubic. By the *Tschirnhaus transformation*

$$u = t + \tfrac{1}{3}s_1$$

the general cubic polynomial takes the form

$$u^3 + pu + q.$$

If we can find the zeros of this it is an easy matter to find them for the general cubic. Following the above procedure for this polynomial we have explicitly

$$y^3 + z^3 = -27q$$
$$y^3 z^3 = -27p^3.$$

from which it follows that y^3 and z^3 are the zeros of the quadratic polynomial

$$t^2 + 27qt - 27p^3.$$

Quartic. The relevant Tschirnhaus transformation is now

$$u = t + \tfrac{1}{4}s_1$$

which reduces the quartic to the form

$$t^4 + pt^2 + qt + r.$$

It is now the case that in the above procedure

$$y_1 + y_2 + y_3 = 2p$$
$$y_1 y_2 + y_1 y_3 + y_2 y_3 = p^2 - 4r$$
$$y_1 y_2 y_3 = -q^2.$$

The resolvent cubic then takes the form

$$t^3 - 2pt^2 + (p^2 - 4r)t + q^2,$$

its zeros are y_1, y_2, y_3, and we have

$$2t_1 = \sqrt{-y_1} + \sqrt{-y_2} + \sqrt{-y_3},$$
$$2t_2 = \sqrt{-y_1} - \sqrt{-y_2} - \sqrt{-y_3},$$
$$2t_3 = -\sqrt{-y_1} + \sqrt{-y_2} - \sqrt{-y_3},$$
$$2t_4 = -\sqrt{-y_1} - \sqrt{-y_2} + \sqrt{-y_3},$$

the square roots being chosen so that

$$\sqrt{-y_1} \cdot \sqrt{-y_2} \cdot \sqrt{-y_3} = -q.$$

Exercises

15.1 If K is a countable field and $L:K$ is finitely generated show that L is countable. Hence show that $\mathbf{R}:\mathbf{Q}$ and $\mathbf{C}:\mathbf{Q}$ are not finitely generated.

15.2 Calculate the transcendence degrees of the following extensions:
(a) $\mathbf{Q}(t, u, v, w):\mathbf{Q}$ where t, u, v, w are independent transcendental elements over $\mathbf{Q}$.
(b) $\mathbf{Q}(t, u, v, w):\mathbf{Q}$ where $t^2 = 2$, u is transcendental

over $\mathbf{Q}(t)$, $v^3 = t + 5$, and w is transcendental over $\mathbf{Q}(t, u, v)$.

(c) $\mathbf{Q}(t, u, v) : \mathbf{Q}$ where $t^2 = u^3 = v^4 = 7$.

15.3 Show that in Lemma 15.1 the degree $[L:M]$ is *not* independent of choice of M. (Hint: consider $K(t^2)$ as a subfield of $K(t)$.)

15.4 Suppose $K \subseteq L \subseteq M$ and each of $M:K$, $L:K$ is finitely generated. Show that $M:K$ and $L:K$ have the same transcendence degree if and only if $M:L$ is finite.

15.5 Using the result that every finite group is isomorphic to a subgroup of $\mathbf{S}_n$ for some n, show that any finite group can be isomorphic to the Galois group of some finite extension. (It is an unsolved problem whether any group can be the Galois group of a finite *normal* extension of $\mathbf{Q}$.)

15.6 For any field K show that $x^3 - 3x + 1$ is either irreducible or splits in K. (Hint: show that any zero is a rational expression in any other zero.)

15.7 Let K be a field of characteristic zero, and suppose that $L:K$ is finite and normal with Galois group G. For any $a \in L$ define the *trace*

$$T(a) = \tau_1(a) + \cdots + \tau_n(a)$$

where $\tau_1, \cdots, \tau_n$ are the distinct elements of G. Show that $T(a) \in K$ and that T is a surjective map $L \to K$.

15.8 If in the previous exercise G is cyclic with generator τ, show that $T(a) = 0$ if and only if $a = b - \tau(b)$ for some $b \in L$.

15.9 Solve by radicals the following polynomials over Q:

(a) $t^3 - 7t + 5$

(b) $t^3 - 7t + 6$

(c) $t^4 + 5t^3 - 2t - 1$

(d) $t^4 + 4t + 2$.

15.10 Show that a finitely generated algebraic extension is finite, and hence find an algebraic extension which is not finitely generated.

15.11 Mark the following true or false.

(a) Every finite extension is finitely generated.

(b) Every finitely generated extension is algebraic.

(c) The transcendence degree of a finitely generated extension is invariant under isomorphism.

(d) If $t_1, \cdots, t_n$ are independent transcendental elements then their elementary symmetric polynomials are also independent transcendental elements.

(e) The Galois group of the general polynomial of degree n is soluble for all n.

(f) The general quintic polynomial is soluble by radicals.

(g) The norm map is a field homomorphism.

(h) The only proper subgroups of $\mathbf{S}_3$ are 1 and $\mathbf{A}_3$.

(i) The general polynomial of degree n 'over' K is a polynomial in one indeterminate over K.

(j) The transcendence degree of $\mathbf{Q}(t):\mathbf{Q}$ is 1.

CHAPTER SIXTEEN

Finite fields

Fields which have finitely many elements play important parts in many branches of mathematics: number theory, group theory, projective geometry, and so on. The most familiar examples of such fields are the fields $\mathbf{Z}_p$ for prime p, but these are not all. In this chapter we give a complete classification of all finite fields. It turns out that a finite field is uniquely determined up to isomorphism by the number of elements it contains; that this number must be a power of a prime; and that for every prime p and integer $n > 0$ there exists a field with p^n elements.

We begin by proving the second of these three statements.

16.1 Theorem. *If F is a finite field then F has characteristic $p > 0$, and the number of elements in F is p^n where n is the degree of F over its prime subfield.*

Proof. Let P be the prime subfield of F. P is not isomorphic to $\mathbf{Q}$ since $\mathbf{Q}$ is infinite, so P is isomorphic to $\mathbf{Z}_p$ for some prime p by 1.2, and hence F has characteristic p. By 4.1 F is a vector space over P. This vector space has finitely many elements, so must be finite-dimensional. Hence $[F:P] = n$ is finite. Let $x_1, \cdots, x_n$ be a basis for F over P. Every element of F is uniquely expressible in the form

$$\lambda_1 x_1 + \cdots + \lambda_n x_n$$

where $\lambda_1, \cdots, \lambda_n \in P$. Each λ_i may be chosen in p ways since $|P| = p$, hence there are p^n such expressions. Therefore $|F| = p^n$.

Thus there do not exist fields with 6, 10, 12, 14, 18, 20 $\cdots$ elements. Notice the contrast with group theory, where there exist groups of any given order. However, there exist non-isomorphic groups with equal orders. To show that this cannot happen for finite fields we must first prove a lemma which is implicit in a result of Chapter 8.

16.2 Lemma. *Let K be a field of characteristic $p > 0$. Then the map $\phi: K \to K$ defined by $\phi(k) = k^p$ ($k \in K$) is a field monomorphism. If K is finite ϕ is an automorphism.*

Proof. Let $x, y \in K$. Then

$$\phi(xy) = (xy)^p = x^p y^p = \phi(x)\phi(y).$$

Also

$$\phi(x+y) = (x+y)^p$$

$$= x^p + px^{p-1}y + \binom{p}{2}x^{p-2}y^2 + \cdots + pxy^{p-1} + y^p$$

by the binomial theorem. As we observed in Chapter 8 when giving an example of an inseparable polynomial, the binomial coefficient $\binom{p}{r}$ is divisible by p if $1 \leq r \leq p-1$. Hence this sum reduces to

$$x^p + y^p$$

$$= \phi(x) + \phi(y).$$

Therefore ϕ is a homomorphism. But $\phi(1) = 1 \neq 0$ so by 3.3 ϕ is a monomorphism.

If K is finite any monomorphism $K \to K$ is automatically surjective by counting elements, so ϕ is an automorphism.

Definition. If K is a field of characteristic $p > 0$ then the map $\phi : K \to K$ defined by $\phi(k) = k^p \, (k \in K)$ is the *Frobenius monomorphism* of K. If K is finite it is referred to as the *Frobenius automorphism*.

For the field $\mathbf{Z}_5$ it turns out that ϕ is the identity map, which is not very inspiring. For the field of Exercise 1.6 we have $\phi(0) = 0$, $\phi(1) = 1$, $\phi(\alpha) = \beta$, $\phi(\beta) = \alpha$, so that ϕ is not always the identity.

16.3 Theorem. *Let p be any prime number and n any integer > 0. A field F has $q = p^n$ elements if and only if it is a splitting field for $f(t) = t^q - t$ over the prime subfield $P \cong \mathbf{Z}_p$ of F.*

Proof. Suppose that $|F| = q$. Now the set $F \backslash \{0\}$ forms a group under multiplication, of order $q - 1$; so that if $0 \neq x \in F$ then $x^{q-1} = 1$. Hence $x^q - x = 0$. Since $0^q - 0 = 0$ every element of F is a zero of $t^q - t$, so that $f(t)$ splits in F. Since the zeros of f exhaust F they certainly generate it, so F is a splitting field for f over P.

Conversely let K be a splitting field for f over Z_p. Since $Df = -1$ which is prime to f, all the zeros of f in K are distinct, and so f has exactly q zeros. Suppose x and y are zeros of f. Then $x^q = \phi^n(x)$ where ϕ is the Frobenius monomorphism. So ϕ^n is also a monomorphism. Therefore

$$(xy)^q - xy = x^q y^q - xy = xy - xy = 0$$

$$(x+y)^q - (x+y) = x^q + y^q - (x+y) = (x+y) - (x+y) = 0$$

$$(x^{-1})^q - x^{-1} = x^{-q} - x^{-1} = x^{-1} - x^{-1} = 0.$$

Hence the set of zeros of f in K is a field, which must therefore be the whole splitting field K. Therefore $|K| = q$.

Since splitting fields exist and are unique up to isomorphism, we deduce:

16.4 Theorem. *A finite field must have $q = p^n$ elements where p is a prime number and n is a positive integer. For each such q there exists up to isomorphism precisely one field with q*

elements, which can be constructed as a splitting field for $t^q - t$ over $\mathbf{Z}_p$.

Notation. The field with q elements is written **GF**(q).

(**GF** stands for *Galois Field* in honour of the discoverer.)

The multiplicative group

The above classification of finite fields, although a useful result in itself, does not give any detailed information on the deeper structure. There are many questions we might ask – what are the subfields? How many are there? What are the Galois groups? We content ourselves with proving one important theorem, which gives the structure of the multiplicative group $F \backslash \{0\}$ of any finite field F. First we need to know a little more about abelian groups.

Definition. The *exponent* $e(G)$ of a finite group G is the least common multiple of the orders of the elements of G.

Clearly $e(G)$ divides the order of G. In general G need not possess an element of order $e(G)$; for example if $G = \mathbf{S}_3$ then $e(G) = 6$ but G has no element of order 6. But abelian groups are better behaved in this respect:

16.5 Lemma. *Any finite abelian group G contains an element of order $e(G)$.*

Proof. Let $e = e(G) = p_1^{\alpha_1} \cdots p_n^{\alpha_n}$ where the p_i are distinct primes and $\alpha_i \geq 1$. Then G must possess elements g_i whose orders are divisible by $p_i^{\alpha_i}$ from the definition of $e(G)$. Then a suitable power a_i of g_i has order $p_i^{\alpha_i}$. Define

$$g = a_1 a_2 \cdots a_n.$$

Suppose that $g^m = 1$ where $m \geq 1$. Then

$$a_i^m = a_1^{-m} \cdots a_{i-1}^{-m} a_{i+1}^{-m} \cdots a_n^{-m}.$$

So, if $q = p_1^{\alpha_1} \cdots p_{i-1}^{\alpha_{i-1}} p_{i+1}^{\alpha_{i+1}} \cdots p_n^{\alpha_n}$ then $a_i^{mq} = 1$. But q is prime to the order of a_i, so $p_i^{\alpha_i}$ divides m. Hence e divides m. But clearly $g^e = 1$. Hence g has order e, which is what we want.

16.6 Corollary. *If G is a finite abelian group such that $e(G) = |G|$ then G is cyclic.*

Proof. The element g constructed above generates G.

We can apply this corollary immediately.

16.7 Theorem. *If G is a finite subgroup of the multiplicative group $K\backslash\{0\}$ of a field K, then G is cyclic.*

Proof. Since multiplication in K is commutative, G is an abelian group. Let $e = e(G)$. Then for any $x \in G$ we have $x^e = 1$, so that x is a zero of the polynomial $t^e - 1$ over K. By 2.8 there are at most e zeros of this polynomial, so that $|G| \le e$. But $e \le |G|$, hence $e = |G|$ and by 16.6 G is cyclic.

16.8 Corollary. *The multiplicative group of a finite field is cyclic.*

Therefore for any finite field F there is at least one element x such that every non-zero element of F is a power of x. We give two examples.

1 The field **GF**(11). The powers of 2, in order, are

$$1, 2, 4, 8, 5, 10, 9, 7, 3, 6, 1$$

so 2 generates the multiplicative group. On the other hand the powers of 4 are

$$1, 4, 5, 9, 3, 1$$

so that 4 does not generate the group.

2 The field **GF**(25). This can be constructed as a splitting field for $t^2 - 2$ over $\mathbf{Z}_5$ since $t^2 - 2$ is irreducible and of degree 2. We can therefore represent the elements of **GF**(25) in the form $a + b\alpha$ where $\alpha^2 = 2$. There is no harm in writing $\alpha = \sqrt{2}$. By trial and error we are led to consider the element $2 + \sqrt{2}$. Successive powers of this are

$$1, 2+\sqrt{2}, 1+4\sqrt{2}, 4\sqrt{2}, 3+3\sqrt{2}, 2+4\sqrt{2}, 2, 4+2\sqrt{2},$$
$$2+3\sqrt{2}, 3\sqrt{2}, 1+\sqrt{2}, 4+3\sqrt{2}, 4, 3+4\sqrt{2}, 4+\sqrt{2}, \sqrt{2},$$
$$2+2\sqrt{2}, 3+\sqrt{2}, 3, 1+3\sqrt{2}, 3+2\sqrt{2}, 2\sqrt{2},$$
$$4+4\sqrt{2}, 1+2\sqrt{2}, 1.$$

Hence $2 + \sqrt{2}$ generates the multiplicative group.

There is no known procedure for finding a generator other than enlightened trial and error. Fortunately the existence of a generator is usually sufficient information.

Exercises

16.1   For which of the following values of n does there exist a field with n elements?

$$1, 2, 3, 4, 5, 6, 17, 24, 312, 65536, 65537, 83521,$$
$$103823, 2^{19937} - 1.$$

16.2   What do the results of this chapter imply with regard to Exercises 8.3 and 8.4?

16.3   Construct fields having 8, 9, and 16 elements.

16.4   Let ϕ be the Frobenius automorphism of **GF**(p^n). Find the least value of $m > 0$ such that ϕ^m is the identity map.

16.5   Show that the subfields of **GF**(p^n) are isomorphic to **GF**(p^r) where r divides n; and there exists a unique subfield for each such r.

16.6   Show that the Galois group of **GF**(p^n):**GF**(p) is cyclic of order n, generated by the Frobenius auto-morphism ϕ. Show that for finite fields the Galois correspondence is a bijection, and find the Galois groups of

$$\mathbf{GF}(p^n):\mathbf{GF}(p^m)$$

whenever m divides n.

16.7 Are there any composite numbers r which always divide the binomial coefficient $\binom{r}{s}$ for $1 \leq s \leq r-1$?

16.8 Find generators for the multiplicative groups of $\mathbf{GF}(n)$ when $n = 8, 9, 13, 17, 19, 23, 29, 31, 37, 41$, or 49.

16.9 Show that the additive group of $\mathbf{GF}(p^n)$ is a direct product of n cyclic groups of order p.

16.10 By considering the field $\mathbf{Z}_2(t)$ show that the Frobenius monomorphism is not always an automorphism.

16.11 For which values of n does $\mathbf{S}_n$ contain an element of order $e(\mathbf{S}_n)$? (Hint: use the cycle decomposition to estimate the maximum order of an element of $\mathbf{S}_n$ and compare this with an estimate of $e(\mathbf{S}_n)$.)

16.12 Mark the following true or false.
  (a) There is a finite field with 124 elements.
  (b) There is a finite field whose multiplicative group has 124 elements.
  (c) There is a finite field with 125 elements.
  (d) The multiplicative group of $\mathbf{GF}(19)$ contains an element of order 3.
  (e) All fields with 121 elements are isomorphic.
  (f) $\mathbf{GF}(2401)$ has a subfield isomorphic to $\mathbf{GF}(49)$.
  (g) Any monomorphism from a finite field to itself is an automorphism.

(h) The additive group of a finite field is cyclic.

(i) The multiplicative group of any field is cyclic.

(j) The exponent of a group is the maximum order of its elements.

CHAPTER SEVENTEEN

Regular polygons

We return with more sophisticated weapons to the problems of ruler-and-compass construction. We shall consider the following question: *for which values of n can the regular n-sided polygon be constructed by ruler and compasses?*

The ancient Greeks knew of constructions for 3-, 5-, and 15-gons; and also knew how to construct a 2n-gon given an n-gon, by the obvious method of bisecting the angles. For about 2000 years little progress was made beyond the Greeks. Then, on 30 March 1796, Gauss made the remarkable discovery that the regular 17-gon could be constructed. He was 19 years old at the time. So pleased was he with this discovery that he resolved to dedicate the rest of his life to mathematics, having until then been unable to decide between that and philology. In his *Disquisitiones Arithmeticae* [39] he stated necessary and sufficient conditions for constructibility of the regular n-gon, and proved their sufficiency; he claimed to have a proof of necessity although he never published it. Doubtless he did: Gauss knew a proof when he saw one.

Which constructions are possible?

In order to obtain necessary *and* sufficient conditions for the existence of a ruler-and-compass construction we must prove a more detailed theorem than 5.2. This requires a careful examination of what constructions are possible.

17.1 Lemma. *If P is a subset of $\mathbf{R}^2$ containing the points $(0, 0)$ and $(1, 0)$ then the point (x, y) can be constructed from P whenever x and y lie in the subfield of $\mathbf{R}$ generated by the coordinates of points in P.*

Proof. Given any point (x_0, y_0) it is obvious how to construct $(0, x_0)$ and $(0, y_0)$. From $(0, 0)$ and $(1, 0)$ we can construct the coordinate axes, and then proceed as shown in Fig. 9.

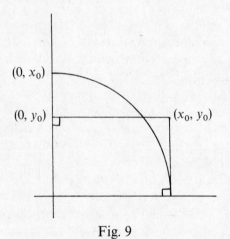

Fig. 9

If we are given $(0, x_0)$ and $(0, y_0)$ the same construction in reverse gives (x_0, y_0). Thus to prove the lemma it is sufficient to show that given $(0, x)$ and $(0, y)$ we can construct $(0, x+y)$, $(0, x-y)$, $(0, xy)$, and $(0, x/y)$ when $y \neq 0$. The first two are obvious; if we swing arcs of radius y centre $(0, x)$ they cut the y-axis at $(0, x+y)$ and $(0, x-y)$. For the other two points we proceed as follows:

Join $(1, 0)$ to $(0, y)$ and draw a line parallel to this through $(0, x)$. This line cuts the x-axis at $(u, 0)$. By similar triangles $u/x = 1/y$, so that $u = x/y$. Taking $x = 1$ (the point $(0, 1)$ is clearly constructible) we can construct $(1/y, 0)$, hence

$(0, 1/y)$; by taking $1/y$ instead of y we get $(xy, 0)$. From these we can find $(0, xy)$ and $(0, x/y)$. See Fig. 10.

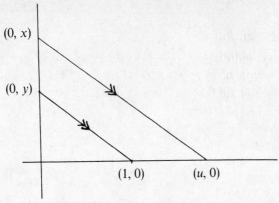

Fig. 10

17.2 Lemma. *Suppose that* $K(\alpha):K$ *is an extension of degree* 2 *such that* $K(\alpha) \subseteq \mathbf{R}$. *Then any point* (z, t) *of* $\mathbf{R}^2$ *whose coordinates* $z, t \in K(\alpha)$ *can be constructed from some suitable finite set of points whose coordinates lie in* K.

Proof. We have $\alpha^2 + p\alpha + q = 0$, where $p, q \in K$. Hence

$$\alpha = \frac{-p \pm \sqrt{(p^2 - 4q)}}{2}$$

and since $K(\alpha) \subseteq R$ then $p^2 - 4q$ must be positive. Using 17.1 the result will follow if we can construct $(0, \sqrt{k})$ for any positive $k \in K$, from finitely many points (x_r, y_r) where $x_r, y_r \in K$.

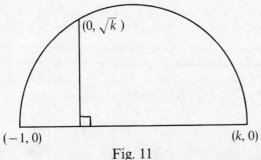

Fig. 11

Construct $(-1, 0)$ and $(k, 0)$. Draw the semicircle with these points as the ends of a diameter, meeting the y-axis at $(0, v)$. By the intersecting chords theorem $v^2 = 1.k$ so that $v = \sqrt{k}$. See Fig. 11.

17.3 Theorem. *Suppose that K is a subfield of $\mathbf{R}$, generated by the coordinates of points in a subset $P \subseteq \mathbf{R}^2$. Let α, β lie in an extension L of K, contained in $\mathbf{R}$, such that there exists a finite series of subfields*

$$K = K_0 \subseteq K_1 \subseteq \cdots \subseteq K_r = L$$

such that

$$[K_{i+1}:K_i] = 2$$

for $i = 0, \cdots, r-1$. Then (α, β) is constructible from P.

Proof. We use induction on r. The case $r = 0$ is covered by 17.1. Otherwise, (α, β) is constructible from finitely many points whose coordinates lie in K_{r-1} by 17.2. By induction these points are constructible from P, so that (α, β) is constructible from P.

From the proof of 5.2 it follows that the existence of such fields K_i is also a necessary condition for (α, β) to be constructible from P.

There is a more useful, but weaker, version of 17.3.

17.4 Proposition. *If K is a subfield of $\mathbf{R}$, generated by the coordinates of points in a subset $P \subseteq \mathbf{R}^2$, and if α and β lie in a normal extension L of K such that $L \subseteq \mathbf{R}$ and $[L:K] = 2^r$ for some integer r, then (α, β) is constructible from P.*

Proof. $L:K$ is separable since the characteristic is zero. Let G be the Galois group of $L:K$. By 11.1 part (1) $|G| = 2^r$ so that G is a 2-group. By 13.10 G has a series of normal subgroups

$$1 = G_0 \subseteq G_1 \subseteq \cdots \subseteq G_r = G$$

such that $|G_i| = 2^i$. Let K_i be the fixed field $G_{r-i}^\dagger$. Then by

11.1 part (3) $[K_{j+1}:K_j] = 2$ for all j. By 17.3 (α, β) is constructible from P.

Regular polygons

We shall use a mixture of algebraic and geometric ideas to find those values of n for which the regular n-gon is constructible. To save breath, let us make the following (non-standard):

Definition. The positive integer n is *constructive* if the regular n-gon is constructible by ruler and compasses.

The first step is to reduce the problem to *prime-power* values of n.

17.5 Lemma. *If n is constructive and m divides n then m is constructive. If m and n are coprime and constructive then mn is constructive.*

Proof. If m divides n then we can construct a regular m-gon by joining every dth vertex of a regular n-gon where $d = n/m$.

If m and n are coprime then there exist integers a, b such that $am + bn = 1$. Therefore

$$\frac{1}{mn} = a \cdot \frac{1}{n} + b \cdot \frac{1}{m}.$$

Hence from angles $2\pi/m$ and $2\pi/n$ we can construct $2\pi/mn$, and from this we obtain a regular mn-gon.

From this we immediately have:

17.6 Corollary. *Suppose that $n = p_1^{\alpha_1} \cdots p_r^{\alpha_r}$ where $p_1, \cdots, p_r$ are distinct primes. Then n is constructive if and only if each $p_i^{\alpha_i}$ is constructive.*

Another obvious result:

17.7 Lemma. *For any positive integer* α *the number* 2^{α} *is constructive.*

Proof. The angle can be bisected by ruler and compasses, and the result follows by induction on α.

This reduces the problem to considering *odd* prime-powers. Now we bring in the algebra. In the complex plane the set of nth roots of unity forms the vertices of a regular n-gon. Further, these roots of unity are the zeros in **C** of the polynomial

$$t^n - 1 = (t-1)(t^{n-1} + t^{n-2} + \cdots + t + 1).$$

We concentrate on the second factor on the right hand side.

17.8 Lemma. *Let p be a prime such that p^n is constructive. Let ζ be a primtive p^nth root of unity in* **C**. *Then the degree of the minimum polynomial of ζ over* **Q** *is a power of 2.*

Proof. We take $\zeta = e^{2\pi i/p^n}$. Since p^n is constructive we can construct the point (α, β) where $\alpha = \cos(2\pi/p^n)$ and $\beta = \sin(2\pi/p_n)$ by projecting a vertex of the regular p^n-gon on to the coordinate axes. Hence by 5.2 we have

$$[\mathbf{Q}(\alpha, \beta) : \mathbf{Q}] = 2^r$$

for some integer r. Therefore

$$[\mathbf{Q}(\alpha, \beta, i) : \mathbf{Q}] = 2^{r+1}$$

But $\mathbf{Q}(\alpha, \beta, i)$ contains $\alpha + i\beta = \zeta$, so that $[\mathbf{Q}(\zeta) : \mathbf{Q}]$ is a power of 2, since $\mathbf{Q}(\zeta) \subseteq \mathbf{Q}(\alpha, \beta, i)$. Hence the degree of the minimum polynomial of ζ over **Q** is a power of 2.

The next step is to calculate the relevant minimum polynomials to find their degrees. It turns out to be sufficient to consider p and p^2 only.

17.9 Lemma. *If p is a prime and ζ is a primitive pth root of unity in* **C** *then the minimum polynomial of ζ over* **Q** *is*

$$f(t) = 1 + t + \cdots + t^{p-1}.$$

Proof. Note that $f(t) = (t^p - 1)/(t - 1)$. We know that $f(\zeta) = 0$ since $\zeta^p - 1 = 0$ and $\zeta \neq 1$. We are home if we can show that $f(t)$ is irreducible. Put $t = 1 + u$ where u is a new indeterminate. Then $f(t)$ is irreducible over **Q** if and only if $f(1 + u)$ is irreducible. But

$$f(1 + u) = \frac{(1 + u)^p = 1}{u}$$

$$= u^{p-1} + p \cdot h(u)$$

where h is a polynomial in u over **Z** with constant term 1, by the usual remark about binomial coefficients. By Eisenstein's criterion 2.5 $f(1 + u)$ is irreducible over **Q**.

17.10 Lemma. *If p is a prime and ζ is a primitive p^2th root of unity in **C** then the minimum polynomial of ζ over **Q** is*

$$g(t) = 1 + t^p + \cdots + t^{p(p-1)}.$$

Proof. Note that $g(t) = (t^{p^2} - 1)/(t^p - 1)$. Now $\zeta^{p^2} - 1 = 0$ but $\zeta^p - 1 \neq 0$ so that $g(\zeta) = 0$. It suffices to show that $g(t)$ is irreducible over **Q**. As before we make the substitution $t = 1 + u$. Then

$$g(1 + u) = \frac{(1 + u)^{p^2} - 1}{(1 + u)^p - 1}$$

and modulo p this is

$$\frac{(1 + u^{p^2}) - 1}{(1 + u^p) - 1}$$

$$= u^{p(p-1)}.$$

Therefore $g(1 + u) = u^{p(p-1)} + p \cdot k(u)$ where k is a polynomial in u over **Z**. From the alternative expression

$$g(1 + u) = 1 + (1 + u)^p + \cdots + (1 + u)^{p(p-1)}$$

it follows that k has constant term 1. By Eisenstein's criterion 2.5 $g(1 + u)$ is irreducible over **Q**.

We now come to the main result.

17.11 Theorem (*Gauss*). *The regular n-gon is constructible by ruler and compasses if and only if*

$$n = 2^r p_1 \cdots p_s$$

where r and s are integers ≥ 0, *and* $p_1, \cdots, p_s$ *are odd primes of the form*

$$p_i = 2^{2^{r_i}} + 1$$

for positive integers r_i.

Proof. Let n be constructive. Then $n = 2^r p_1^{\alpha_1} \cdots p_s^{\alpha_s}$ where $p_1, \cdots, p_s$ are distinct odd primes. By 17.6 each $p_i^{\alpha_i}$ is constructive. If $\alpha_i \geq 2$ then p_i^2 is constructive by 17.4. Hence the degree of the minimum polynomial of a primitive p_i^2th root of unity over $\mathbf{Q}$ is a power of 2 by 17.8. By 17.10 $p_i(p_i - 1)$ is a power of 2, which cannot happen since p_i is odd. Therefore $\alpha_i = 1$ for all i.

Therefore p_i is constructive. By 17.9

$$p_i - 1 = 2^{s_i}$$

for suitable s_i. Suppose that s_i has an odd divisor $a > 1$, so that $s_i = ab$. Then

$$p_i = (2^b)^a + 1$$

which is divisible by $2^b + 1$ since

$$t^a + 1 = (t+1)(t^{a-1} - t^{a-2} + \cdots + 1)$$

when a is odd. So p_i cannot be prime. Hence s_i has no odd factors, so

$$s_i = 2^{r_i}$$

for some $r_i > 0$.

This establishes the necessity of the given form of n. Now we prove sufficiency. By 17.6 we need consider only prime-power factors of n. By 17.7 2^r is constructive. We must show that each p_i is constructive. Let ζ be a primitive p_ith root of unity. Then $[\mathbf{Q}(\zeta):\mathbf{Q}] = p_i - 1 = 2^a$ for some a by 17.9. Now $\mathbf{Q}(\zeta)$ is a splitting field for $f(t) = 1 + \cdots + t^{p-1}$ over $\mathbf{Q}$, so that $\mathbf{Q}(\zeta):\mathbf{Q}$ is normal. It is also separable since the characteristic is zero. By 14.4 the Galois group $\Gamma(\mathbf{Q}(\zeta):\mathbf{Q})$ is abelian. Let $K = \mathbf{R} \cap \mathbf{Q}(\zeta)$. Then $\cos(2\pi/p_i) = (\zeta + \zeta^{-1})/2$ $\in K$. Now $\mathbf{Q}(\zeta):K$ has degree 2 so by 11.1 $\Gamma(\mathbf{Q}(\zeta):K)$ is a subgroup of $G = \Gamma(\mathbf{Q}(\zeta):\mathbf{Q})$ of order 2 and further it is a normal subgroup since G is abelian. Therefore $K:\mathbf{Q}$ is a normal extension of degree 2^{a-1}. Therefore by 17.4 the point $(0, \cos(2\pi/p_i))$ is constructible. Hence p_i is constructive, and the proof is complete.

Fermat numbers

The problem now reduces to number theory. If we make the:

Definition. The nth *Fermat number* is $F_n = 2^{2^n} + 1$

then the question becomes: *when is F_n prime?*

In 1640 Pierre de Fermat noticed that $F_0 = 3$, $F_1 = 5$, $F_2 = 17$, $F_3 = 257$, and $F_4 = 65537$ are all prime. He conjectured that F_n is prime for all n, but this was disproved by Euler in 1732. The latest figures for Fermat numbers (at the time of writing) are shown in the accompanying table. (The bracketed initials after discoverer's names indicate their faithful electronic computers, without which such mathematical big-game hunting would be a desperate task. Those numbers marked 'composite' in the table but without factors listed are those which have been shown composite by a test which does not give the factors. The factors are listed in the form $k2^m + 1$ where k is odd, a convenient form in which to express the large numbers involved.)

It will be seen that the only known Fermat *primes* are those found by Fermat himself.

Inspection of the table reveals:

17.12 Proposition. *The only primes $p < 10^{40000}$ for which the regular p-gon is constructible are* 2, 3, 5, 17, 257, 65537.

Known factors of Fermat numbers

n	Factors $k2^m + 1$ of F_n		Date	Discoverer
	k	m		
0	prime		1640	Fermat
1	prime		1640	Fermat
2	prime		1640	Fermat
3	prime		1640	Fermat
4	prime		1640	Fermat
5	5	7	1732	Euler
	52347	7	1732	Euler
6	composite		1878	Lucas
	1071	8	1880	Landry
	262814145745	8	1880	Landry, Le Lasseur
7	composite		1905	Morehead, Western
	116503103764643	9	1970	Brillhart, Morrison (IBM 360–91)
	11141971095088142685	9	1970	Brillhart, Morrison (IBM 360–91)
8	composite		1909	Morehead, Western
9	37	16	1903	Western
10	11131	12	1953	Selfridge (SWAC)
	395937	14	1962	Brillhart
11	39	13	1899	Cunningham
	119	13	1899	Cunningham
12	7	14	1877	Lucas, Pervušin
	397	16	1903	Western
	973	16	1903	Western
13	composite		1961	Paxson (IBM 7090)
14	composite		1963	Paxson, Hurwitz, Selfridge (IBM 7090)
15	579	21	1925	Kraitchik
16	1575	19	1953	Selfridge (SWAC)
18	13	20	1903	Western
23	5	25	1878	Pervušin
36	5	39	1886	Seelhoff
38	3	41	1903	Cullen, Cunningham, Western
39	21	41	1956	Robinson (SWAC)
55	29	57	1956	Robinson (SWAC)
58	95	61	1957	Robinson (SWAC)
63	9	67	1956	Robinson (SWAC)
73	5	75	1906	Morehead
77	425	79	1957	Robinson, Selfridge (SWAC)
81	271	84	1957	Robinson, Selfridge (SWAC)

Known factors of Fermat numbers

	Factors $k2^m + 1$ of F_n		Date	Discoverer
n	k	m		
117	7	120	1956	Robinson (SWAC)
125	5	127	1956	Robinson (SWAC)
144	17	147	1956	Robinson (SWAC)
150	1575	157	1956	Robinson (SWAC)
207	3	209	1956	Robinson (SWAC)
226	15	229	1956	Robinson (SWAC)
228	29	231	1956	Robinson (SWAC)
250	403	252	1957	Robinson, Selfridge (SWAC)
267	177	271	1957	Robinson, Selfridge (SWAC)
268	21	276	1956	Robinson (SWAC)
284	7	290	1956	Robinson (SWAC)
316	7	320	1956	Robinson (SWAC)
452	27	455	1956	Robinson (SWAC)
1945	5	1947	1957	Robinson (SWAC)

Constructions

Many constructions for the 17-gon have been devised, the earliest published being that of Huguenin (see Klein [41]) in 1803. For several of these constructions there are proofs of their correctness which use only synthetic geometry (i.e. ordinary Euclidean geometry without coordinates). A series of papers giving a construction for the regular 257-gon was published by Richelot [46] in 1832, under one of the longest titles I have ever seen. In [28] Bell tells of an over-zealous research student being sent away to find a construction for the 65537-gon, and reappearing with one 20 years later. This story, though apocryphal, is not far from the truth; Professor Hermes of Lingen spent 10 years on the problem, and his manuscripts are still (I believe) preserved at Göttingen. Whether such monumental labour will ever have much significance is another matter . . . but mathematics is rife with indefatigable calculators. So far no one has actually *performed* a construction for the 65537-gon.

One way to construct a regular 17-gon is to follow faithfully the above theory, which in fact provides a perfectly definite construction after a little extra calculation. With a little ingenuity it is possible to shorten the work. We shall give a construction; the procedure used is taken from Hardy and Wright [13].

Our immediate object is to find radical expressions for the zeros of the polynomial

$$\frac{t^{17}-1}{t-1} = t^{16}+\cdots+t+1 \tag{1}$$

over **C**. Let

$$\theta = 2\pi/17$$

$$\varepsilon_k = e^{ki\theta} = \cos k\theta + i \sin k\theta.$$

The zeros of (1) in **C** are then $\varepsilon_1, \cdots, \varepsilon_{16}$.

The powers of 3 reduced mod 17 are as follows:

m	0	1	2	3	4	5	6	7	8	9	10	11	12	13	14	15
3^m	1	3	9	10	13	5	15	11	16	14	8	7	4	12	2	6 .

We define

$$x_1 = \varepsilon_1 + \varepsilon_9 + \varepsilon_{13} + \varepsilon_{15} + \varepsilon_{16} + \varepsilon_8 + \varepsilon_4 + \varepsilon_2,$$

$$x_2 = \varepsilon_3 + \varepsilon_{10} + \varepsilon_5 + \varepsilon_{11} + \varepsilon_{14} + \varepsilon_7 + \varepsilon_{12} + \varepsilon_6,$$

$$y_1 = \varepsilon_1 + \varepsilon_{13} + \varepsilon_{16} + \varepsilon_4$$

$$y_2 = \varepsilon_9 + \varepsilon_{15} + \varepsilon_8 + \varepsilon_2$$

$$y_3 = \varepsilon_3 + \varepsilon_5 + \varepsilon_{14} + \varepsilon_{12}$$

$$y_4 = \varepsilon_{10} + \varepsilon_{11} + \varepsilon_7 + \varepsilon_6.$$

Now

$$\varepsilon_k + \varepsilon_{17-k} = 2\cos k\theta \tag{2}$$

for $k = 1, \cdots, 16$, so that

$$x_1 = 2(\cos\theta + \cos 8\theta + \cos 4\theta + \cos 2\theta)$$

$$x_2 = 2(\cos 3\theta + \cos 7\theta + \cos 5\theta + \cos 6\theta) \qquad (3)$$

$$y_1 = 2(\cos\theta + \cos 4\theta)$$

$$y_2 = 2(\cos 8\theta + \cos 2\theta)$$

$$y_3 = 2(\cos 3\theta + \cos 5\theta)$$

$$y_4 = 2(\cos 7\theta + \cos 6\theta).$$

From (1) it follows that

$$x_1 + x_2 = -1.$$

Using (3) and the equation $2\cos m\theta \cos n\theta = \cos(m+n)\theta + \cos(m-n)\theta$ we find that

$$
\begin{aligned}
x_1 x_2 = 2\{ &\cos 4\theta + \cos 2\theta + \cos 8\theta + \cos 6\theta + \cos 4\theta \\
&+ \cos 6\theta + \cos 5\theta + \cos 7\theta + \cos 11\theta + \cos 5\theta \\
&+ \cos 15\theta + \cos\theta + \cos 13\theta + \cos 3\theta + \cos 14\theta \\
&+ \cos 2\theta + \cos 7\theta + \cos\theta + \cos 3\theta + \cos 11\theta \\
&+ \cos\theta + \cos 9\theta + \cos 2\theta + \cos 10\theta + \cos 5\theta \\
&+ \cos\theta + \cos 9\theta + \cos 5\theta + \cos 7\theta + \cos 3\theta \\
&+ \cos 4\theta + \cos 8\theta \}
\end{aligned}
$$

$$= -4$$

using (2). Hence x_1 and x_2 are zeros of the quadratic polynomial

$$t^2 + t - 4. \qquad (4)$$

Further, $x_1 > 0$ so that $x_1 > x_2$. By trigonometric expansions similar to that above, we find that

$$y_1 + y_2 = x_1$$

$$y_1 y_2 = -1$$

and y_1, y_2 are the zeros of

$$t^2 - x_1 t - 1. \qquad (5)$$

Further, $y_1 > y_2$. Similarly, y_3 and y_4 are the zeros of

$$t^2 - x_2 t - 1 \tag{6}$$

and $y_3 > y_4$.

Now

$$2 \cos \theta + 2 \cos 4\theta = y_1$$

$$4 \cos \theta \cos 4\theta = 2(\cos 5\theta + \cos 3\theta) = y_3$$

so that

$$z_1 = 2 \cos \theta, \quad z_2 = 2 \cos 4\theta$$

are the zeros of

$$t^2 - y_1 t + y_3 \tag{7}$$

and $z_1 > z_2$.

Solving the series of quadratics (4)(5)(6)(7) and using the inequalities to decide which zero is which we obtain the equation

$$\cos \theta =$$

$$= \frac{1}{16}\left\{ -1 + \sqrt{17} + \sqrt{34 - 2\sqrt{17}} + \right.$$

$$\left. + \sqrt{ 68 + 12\sqrt{17} - 16\sqrt{34 + 2\sqrt{17}} - 2(1 - \sqrt{17})\sqrt{34 - 2\sqrt{17}} } \right\}$$

where the square roots are the positive ones.

From this we can deduce a geometrical construction for the 17-gon by constructing the relevant square roots. By using greater ingenuity it is possible to obtain an aesthetically more satisfying construction. The following method is due to Richmond [47].

Let ϕ be the smallest positive acute angle such that $\tan 4\phi = 4$. Then ϕ, 2ϕ, and 4ϕ are all acute. Expression (4) above can be written

$$t^2 + 4t \cot 4\phi - 4$$

whose zeros are
$$2\tan 2\phi, \ -2\cot 2\phi.$$
Hence
$$x_1 = 2\tan 2\phi, \quad x_2 = -2\cot 2\phi.$$
From this it follows that

$$y_1 = \tan\left(\phi+\frac{\pi}{4}\right)$$

$$y_2 = \tan\left(\phi-\frac{\pi}{4}\right)$$

$$y_3 = \tan\phi$$

$$y_4 = -\cot\phi.$$

Then

$$2(\cos 3\theta + \cos 5\theta) = \tan\phi$$

$$4\cos 3\theta \cos 5\theta = \tan\left(\phi-\frac{\pi}{4}\right). \tag{8}$$

Now (as in Fig. 12) let OA, OB be two perpendicular radii of a circle. Make $OI = \frac{1}{4}OB$ and $\angle OIE = \frac{1}{4}\angle OIA$. Find

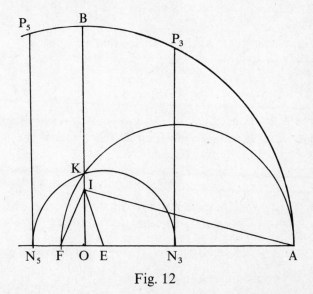

Fig. 12

F on AO produced to make $\angle EIF = \frac{\pi}{4}$. Let the circle on AF as diameter cut OB in K, and let the circle centre E through K cut OA in N_3 and N_5 as shown. Draw N_3P_3 and N_5P_5 perpendicular to OA.

Then $\angle OIA = 4\phi$ and $\angle OIE = \phi$. Also

$$2(\cos \angle AOP_3 + \cos \angle AOP_5) = 2\frac{ON_3 - ON_5}{OA}$$

$$= 4\frac{OE}{OA} = \frac{OE}{OI} = \tan \phi$$

and

$$4 \cos \angle AOP_3 \cos \angle AOP_5 = -4\frac{ON_3 . ON_5}{OA . OA}$$

$$= -4\frac{OK^2}{OA^2}$$

$$= -4\frac{OF}{OA}$$

$$= -\frac{OF}{OI} = \tan\left(\phi - \frac{\pi}{4}\right).$$

Comparing these with (8) we see that

$$\angle AOP_3 = 3\theta$$

$$\angle AOP_5 = 5\theta.$$

Hence A, P_3, P_5 are the 0th, 3rd, and 5th vertices of a regular 17-gon inscribed in the given circle. The other vertices are now easily found.

Exercises

17.1 Verify the following approximate constructions for regular n-gons (due to Oldroyd [23]).

(a) 7-gon. Construct $\cos^{-1}\left(\frac{4+\sqrt{5}}{10}\right)$ giving an angle of approximately $2\pi/7$.

(b) 9-gon. Construct $\cos^{-1}\left(\dfrac{5\sqrt{3}-1}{10}\right)$.

(c) 11-gon. Construct $\cos^{-1}(\frac{8}{9})$ and $\cos^{-1}(\frac{1}{2})$ and take their difference.

(d) 13-gon. Construct $\tan^{-1}(1)$ and $\tan^{-1}\dfrac{4+\sqrt{5}}{20}$ and take their difference.

17.2 Show that for n *odd* the only known constructible n-gons are those for which n is a divisor of $2^{32}-1 = 4294967295$.

17.3 Work out the approximate size of F_{1945}. Explain why it is no easy task to find factors of Fermat numbers.

17.4 Use the equations $641 = 5^4+2^4 = 5.2^7+1$ to show that 641 divides F_5.

17.5 Show that
$$F_{n+1} = 2+F_n F_{n-1} \cdots F_0$$
and deduce that if $m \neq n$ then F_m and F_n are coprime. Hence show that there are infinitely many prime numbers.

17.6 List the values of $n \leq 100$ for which the regular n-gon can be constructed by ruler and compasses.

17.7 Verify the following construction for the regular pentagon.

Draw a circle centre O with two perpendicular radii OP_0, OB. Let D be the midpoint of OB, join $P_0 D$. Bisect $\angle ODP_0$ cutting OP_0 at N. Draw NP_1 perpendicular to OP_0 cutting the circle at P_1. Then P_0 and P_1 are the 0th and 1st vertices of a regular pentagon inscribed in the circle.

17.8 Mark the following true or false.

(a) $2^n + 1$ cannot be prime unless n is a power of 2.

(b) If n is a power of 2 then $2^n + 1$ is always prime.

(c) The regular 771-gon is constructible using ruler and compasses.

(d) The regular 768-gon is constructible using ruler and compasses.

(e) The regular 25-gon is constructible using ruler and compasses.

(f) For an odd prime p the regular p^2-gon is never constructible using ruler and compasses.

(g) If n is an integer > 0 then a line of length $\sqrt{n}$ can always be constructed using ruler and compasses.

(h) A point whose coordinates lie in a normal extension of $\mathbf{Q}$ of 2-power degree is constructible using ruler and compasses.

(i) The regular 51-gon is constructible using ruler and compasses.

(j) If p is a prime then $t^{p^2} - 1$ is irreducible over $\mathbf{Q}$.

CHAPTER EIGHTEEN

The 'Fundamental Theorem of Algebra'

A field is *algebraically closed* if every polynomial over it splits. The result referred to in the title of this chapter states that the field of complex numbers is algebraically closed. This theorem is not fundamental to modern algebra, nor is it entirely a theorem of algebra at all. Hence the inverted commas.

The first proof was given by Gauss in his doctoral dissertation of 1799 under the title (in Latin) *A new proof that every rational integral function of one variable can be resolved into real factors of the first or second degree.* Gauss was being polite in using the word 'new'; his was the first genuine *proof.* Even his proof, from the modern viewpoint, has gaps; but these are topological in nature and not hard to fill. In Gauss's day they were not considered to be gaps at all. Gauss's proof can be found in Hardy [12] p. 492.

There are other proofs. One of a more recondite nature uses complex variable theory (see Titchmarsh [26] p. 118) and is probably the proof most commonly given. There is a proof by Clifford [34] p. 20 which is almost entirely algebraic; the idea is to show that any irreducible polynomial over **R** is of degree 1 or 2. The proof we shall give here is essentially due to Legendre; but his original proof had gaps which we shall fill by using Galois theory.

It is unreasonable to ask for a purely algebraic proof of

the theorem, since the real numbers (and hence the complex numbers) are defined in terms of analytic concepts such as Cauchy sequences, Dedekind cuts, or completeness in an ordering.

Definition. An *ordered field* is a field K with a relation $\leq$ such that:

(1) $k \leq k$ for all $k \in K$.
(2) $k \leq l$ and $l \leq m$ implies $k \leq m$ for all $k, l, m \in K$.
(3) $k \leq l$ and $l \leq k$ implies $k = l$ for all $k, l \in K$.
(4) If $k, l \in K$ then either $k \leq l$ or $l \leq k$.
(5) If $k, l, m \in K$ and $k \leq l$ then $k+m \leq l+m$.
(6) If $k, l, m \in K$ and $k \leq l$ and $0 \leq m$ then $km \leq lm$.

The relation $\leq$ is an *ordering* on K. The relations $<$, $\geq$, $>$ are defined in the obvious way, as are the concepts *positive* and *negative*.

Examples of ordered fields are **Q** and **R**.
We need two simple consequences of the definition of an ordered field.

18.1 Lemma. *Let K be an ordered field. For any $k \in K$ we have $k^2 \geq 0$. The characteristic of K must be zero.*

Proof. If $k \geq 0$ then $k^2 \geq 0$ by (6). So by (3) we may assume $k < 0$. If now we had $-k < 0$ it would follow that

$$0 = k+(-k) < k+0 = k,$$

a contradiction. So $-k \geq 0$, whence $k^2 = (-k)^2 \geq 0$. This proves the first statement.

Therefore $1 = 1^2 > 0$, and hence for any finite n the number

$$n.1 = 1+\cdots+1 > 0$$

so that $n.1 \neq 0$ and K must have characteristic 0.

We quote the following properties of **R**.

18.2 Lemma. R, *with the usual ordering, is an ordered field. Every positive element of* **R** *has a square root in* **R**. *Every odd degree polynomial over* **R** *has a zero in* **R**.

These are all proved in any course in analysis, and depend mainly on the fact that a polynomial function on **R** is continuous.

Our next step is to set up some more Galois-theoretic machinery.

18.3 Lemma. *Let K be a field of characteristic zero, such that for some prime p every finite extension M of K with M ≠ K has* $[M:K]$ *divisible by p. Then every finite extension of K has degree a power of p.*

Proof. Let N be an extension of K. The characteristic is zero so $N:K$ is separable. By passing to a normal closure we may assume $N:K$ is also normal, so that the Galois correspondence is bijective. Let G be the Galois group of $N:K$, and let P be a Sylow p-subgroup of G. The fixed field $P^\dagger$ has degree $[P^\dagger:K]$ equal to the index of P in G (11.1 part (3)), but this is prime to P. By hypothesis therefore $P^\dagger = K$ so that $P = G$. Then $[N:K] = |G| = p^n$ for some n.

18.4 Theorem. *Let K be an ordered field in which every positive element has a square root and every odd-degree polynomial has a zero. Then K(i) is algebraically closed, where* $i^2 = -1$.

Proof. K cannot have any extensions of finite odd degree greater than 1. For suppose $[M:K] = r > 1$ where r is odd. Let $\alpha \in M \backslash K$ have minimum polynomial m. Then ∂m divides r, so is odd. By hypothesis m has a zero in K, so is reducible, contradicting 3.2. Hence every finite extension of K has even degree over K. The characteristic of K is 0 by 18.1, so by 18.3 every finite extension of K has 2-power degree.

Let $M \neq K(i)$ be any finite extension of $K(i)$ where $i^2 = -1$. By taking a normal closure we may assume $M:K$

is normal, so the Galois group of $M:K$ is a 2-group. Using 13.10 and the Galois correspondence we can find an extension N of $K(i)$ of degree $[N:K(i)] = 2$. By the formula for solving quadratic equations, we have

$$N = K(i)(\alpha)$$

where $\alpha^2 \in K(i)$. But if $a, b \in K$ we have

$$\sqrt{a+bi} = \sqrt{\frac{a+\sqrt{a^2+b^2}}{2}} + i\sqrt{\frac{-a+\sqrt{a^2+b^2}}{2}}$$

where the square root of a^2+b^2 is the positive one, and the signs of the other two square roots are chosen to make their product equal to b. The square roots exist in K since the elements inside them are positive, as is easily checked.

Therefore $\alpha \in K$, so that $N = K(i)$, which contradicts our assumption on N. Therefore $M = K(i)$, and $K(i)$ has no finite extensions of degree > 1. Hence any irreducible polynomial over $K(i)$ has degree 1, otherwise a splitting field would have finite degree > 1 over $K(i)$. Therefore $K(i)$ is algebraically closed.

18.5 Corollary (*Fundamental Theorem of Algebra*). *The field* **C** *of complex numbers is algebraically closed.*

Proof. Put **R** $= K$ in 18.4 and use 18.2.

Exercises

18.1   Show that a field K is algebraically closed if and only if $L:K$ algebraic implies $L = K$.

18.2   Show that every algebraic extension of **R** is isomorphic to **R**:**R** or **C**:**R**.

18.3   Prove by transfinite induction that every field K has an algebraically closed extension field L. (Keep adjoining zeros of irreducible polynomials until there are none left.)

18.4   Show that **C**, with the traditional field operations, cannot be given the structure of an ordered field. If we allow different field operations, can the *set* **C** be given the structure of an ordered field?

18.5   Prove the theorem whose statement is the title of Gauss's doctoral dissertation mentioned at the beginning of the chapter.

18.6   Suppose that $K : \mathbf{Q}$ is a finitely generated extension. Prove that there exists a **Q**-monomorphism $K \to \mathbf{C}$. (Hint: Use cardinality considerations to adjoin transcendental elements, and algebraic closure of **C** to adjoin algebraic elements.) Is the theorem true for **R** rather than **C**?

18.7   Let K be an ordered field. A subset S of K is *bounded above* if there exists $k \in K$ such that for all $s \in S$ we have $s \leq k$; and k is an *upper bound*. An element l is a *least upper bound* for S if it is an upper bound, and for any other upper bound m we have $l \leq m$.

   Show that there exist K and S such that S has upper bounds but no least upper bound.

18.8   An ordered field K is *complete* if every bounded subset S of K has a least upper bound. Prove that every complete ordered field is isomorphic to the field R of real numbers, by an order-preserving isomorphism.

   (Hint: K contains a subfield isomorphic to **Q** with the natural ordering. Let T be the set of least upper bounds of bounded subsets of **Q**. Show that T is isomorphic to R, and that T is the whole of K.)

18.9   From the definition of R as a complete ordered field, deduce the properties asserted of it in Lemma 18.2.

18.10 Mark the following true or false.

(a) The field **C** is algebraically closed.

(b) The field **A** of algebraic numbers is algebraically closed.

(c) The field **A** is an ordered field.

(d) There is no ordering on **C** making it into an ordered field.

(e) Every polynomial over **R** splits in **C**.

(f) If f is a polynomial over **R** and K is a splitting field for f then either $K = $ **R** or $K = $ **C**.

(g) Every ordered field has characteristic zero.

(h) Every field of characteristic zero can be ordered.

(i) In an ordered field, every square is positive.

(j) If every element of a field K has a square root in K, then K cannot be an ordered field.

CHAPTER NINETEEN

Later developments

The ideas of Galois theory have had fruitful application in many areas of mathematics. In this chapter we shall briefly survey, without proofs, three variations on the theme: the Galois theories of commutative rings, division rings, and differential equations. We should also mention the recent surge of activity in the classification of finite simple groups, a problem not yet completely solved. We refer the reader to Carter [6], la Chyl [18], and Tits [27] for surveys of these fascinating discoveries.

Galois theory of commutative rings

One of the more reasonable areas to which one might try to generalize Galois theory is that of commutative rings. Now even for fields we have seen that the Galois correspondence is not always bijective, and we therefore expect extra conditions to obtain a workable theory for rings. One possible way of attacking the problem is given by Chase, Harrison, and Rosenberg in [7], and we outline some of their work here. For details, the reader should consult their memoir. They do not assume any of the standard results of Galois theory, so that their methods provide an alternative approach to the standard theory for fields, albeit one which is not suitable for a *first* treatment.

For the rest of this section the word *ring* will mean *commutative ring-with*-1 in the sense of Chapter 1. Subrings will always contain the 1, but ideals need not.

Let us analyse some of the key steps in our previous treatment of Galois theory. Of great importance is the observation that if $L:K$ is an extension then L has a natural structure as vector space over K. The corresponding notion for rings is that of a *module* (see Hartley and Hawkes [14]). If R is a ring then an *R-module* is an abelian group M together with an *action* of R on M, that is, a map $\mu:R \times M \to M$. For convenience we write $\mu(r, m) = rm$ ($r \in R, m \in M$). Further, the action must satisfy the following axioms: for all $x, y \in M$ and $r, s \in R$,

$$r(x+y) = rx+ry$$
$$(r+s)x = rx+sx$$
$$(rs)x = r(sx)$$
$$1x = x.$$

Homomorphisms of R-modules are defined in the obvious way; they are homomorphisms of the underlying additive groups which preserve the action of R.

If R is a subring of S then (exactly as in Theorem 4.1) it follows that S has a natural structure as R-module. Since for fields K a K-module is just a vector space over K it can be seen that this generalization is a sensible one.

Another important property that we used was the existence of a basis in a vector space. Experience suggests the following replacement. Say that an R-module P is *projective* if whenever there are R-modules A and B, together with homomorphisms $\gamma:P \to B$ and $\alpha:A$ *onto* B, then there exists a homomorphism $\pi:P \to A$ such that $\gamma = \alpha\pi$. In diagrammatic form.

commutes.

We remark that when K is a field, every K-module is projective. For suppose that in the above, A, B, P are all vector spaces over K (i.e. K-modules). Take a basis $(p_i)_{i \in I}$ for P. Since α is onto we may choose a_i $(i \in I)$ belonging to A such that

$$\alpha(a_i) = \gamma(p_i) \qquad (i \in I).$$

Define π on basis elements by

$$\pi(p_i) = a_i$$

and extend it to a linear map $P \to A$ as usual. Then $\gamma = \alpha\pi$ as required. In this way the projectiveness condition generalizes the existence of a basis in vector spaces.

Next we need a condition analogous to 'finite-dimensional' for vector spaces. We say that an R-module M is *finitely generated* if there exist finitely many elements $m_1, \cdots, m_n \in M$ such that every element of M can be expressed in the form

$$r_1 m_1 + \cdots + r_n m_n$$

for suitable $r_1, \cdots, r_n \in R$. When R is a field this says that $m_1, \cdots, m_n$ *span* M, as vector space; so that M is finite-dimensional.

Thus, in order to preserve the sort of behaviour exhibited by the vector space structure of a finite field extension, we shall consider *ring extensions* $S:R$ (i.e. injections $R \to S$, where R is if possible identified with its image in S) such that S is a finitely generated projective R-module. However, more hypotheses are needed yet.

The crucial conditions for fields are the assumption of *separability* and *normality*. In Theorems 10.8 and 10.10 we have shown that $L:K$ is a finite separable normal field extension if and only if K is the fixed field of a finite group of automorphisms of L. In somewhat underhand fashion, we take *this* as another necessary condition for rings. (Thus, to derive the standard theory from [7] we still need all of our results up to Chapter 10.) If we have a ring S and a subring R such that R is the *fixed ring*

$$\{r \in S : g(r) = r \text{ for all } g \in G\}$$

of a finite group G of automorphisms of S, we say that $S:R$ is an *extension with Galois group* G. We shall require our extensions to be of this type.

But even now we require one further condition, which may be expressed in the following form: every homomorphism $\phi : S \to S$ (where S is considered as an R-module) has unique expression in the form

$$\phi(x) = s.g(x) \qquad (x \in S) \tag{*}$$

for suitable $s \in S$, $g \in G$. This condition should be compared with the results of Chapter 9. Uniqueness of the expression (*) follows (in the case of field extensions) by 9.1; existence follows from 9.4 by comparing dimensions.

We now make the following definition: $S:R$ is a *Galois extension* with *Galois group* G if

(1) G is a finite group of automorphisms of S,
(2) R is the fixed ring of G,
(3) S is a finitely generated projective R-module,
(4) Condition (*) above holds.

There is still one technical obstacle to be overcome before we can state a generalization of the Fundamental Theorem. An element e of a ring R is an *idempotent* if $e^2 = e$. Thus 0 and 1 are idempotents; but there may be others. We shall restrict our attention to rings where 0, 1 are the only idempotents; in particular integral domains have this property.

We may now state our Fundamental Theorem in a more general setting.

Theorem. *Let $S:R$ be a Galois extension of rings with Galois group G, such that the only idempotents of S are 0 and 1. Then there is an order-reversing bijection between*

(i) *The set of subgroups of G,*
(ii) *The set of rings T lying between R and S with the property that conditions (3) and (4) above hold for the extension $T:R$.*

This bijection associates with each subgroup of G its fixed ring, and with each intermediate ring T (such that conditions (3) and (4) hold) the group of automorphisms of S that fix every element of T.

It is possible to generalize to this setting the part of the Fundamental Theorem concerning normal subgroups of the Galois group.

Apart from sheer generalization for its own sake, and aside from possible applications to ring theory, this more general theorem is interesting in that it provides an analysis of the crucial ideas in the standard theory; it enables us to see just which properties of fields are being used at various stages in the proof.

Galois theory of division rings

A *division ring* (or *skew field* or *sfield*) is a ring whose non-zero elements form a group under multiplication; it is essentially a field in which multiplication is not necessarily commutative. The simplest example of a division ring which is not a field is the set **H** of *quaternions*, discovered independently by Gauss and Hamilton. This consists of all expressions

$$p + qi + rj + sk$$

where $p, q, r, s \in \mathbf{R}$ and the elements i, j, k multiply together according to:

$$i^2 = j^2 = k^2 = -1$$

$$ij = k, jk = i, ki = j$$

$$ji = -k, kj = -i, ik = -j.$$

This multiplication extends to all quaternions using the distributive law; the addition formula is obvious.

Division rings are sufficiently close to fields for one to hope that some analogue of the Galois theory would carry over. In addition, since division rings are less well understood than fields, such a theory might be of considerable importance.

The first difference is that there are two kinds of vector space over a division ring: left and right. Let Δ be a division ring. In defining a *vector space over* Δ we proceed as for a field, obtaining an abelian group V on which the elements of Δ act as 'scalars'. If $v \in V$, $\delta \in \Delta$ and we write the product as δv then it is natural to assume that

$$\delta_1(\delta_2 v) = (\delta_1 \delta_2)v.$$

On the other hand if we write $v\delta$, then the natural assumption is that

$$(v\delta_1)\delta_2 = v(\delta_1 \delta_2).$$

The first gives a *left* vector space, the second a *right* vector space. For fields we have $\delta_1\delta_2 = \delta_2\delta_1$ and the two types are indistinguishable, but for general division rings they need not be the same.

If Δ is a division ring and Γ is a division subring of Δ then Δ has a natural left vector space structure over Γ *and* a natural right vector space structure. Consequently we can define two degrees,

$$[\Delta:\Gamma]_L \quad \text{and} \quad [\Delta:\Gamma]_R.$$

For some time the question whether these two degrees are always equal was unsolved; but in 1961 Cohn [8] showed they are not.

The *centre* Z of a division ring Δ is the set of all elements $z \in \Delta$ such that $zd = dz$ for all $d \in \Delta$. Since $1 \in Z$, Z is not empty, and is a sub*field* of Δ. There is a related concept: if Σ is a division subring of Δ then the *centralizer* of Σ in Δ is the set of all $z \in \Delta$ such that $zs = sz$ for all $s \in \Sigma$. The centralizer is a division subring.

For any $0 \neq a \in \Delta$, the map I_a which takes an element $x \in \Delta$ to axa^{-1} is an automorphism of Δ; the *inner automorphism* induced by a. Given a group G of automorphisms of Δ we can define the *fixed ring* $G^\dagger$ to be the set of all $d \in \Delta$ such that $g(d) = d$ for all $g \in G$. Then $G^\dagger$ is a division subring. Given any division subring Γ of Δ then the *Galois group*

Γ^* of Δ over Γ is the group of all automorphisms of Δ which leave every element of Γ fixed.

Now the centralizer Σ of Γ in Δ is a division subring of Δ containing the centre Z of Δ. Non-zero elements of Σ determine inner automorphisms of Δ which leave every element of Γ fixed (since elements of Σ commute with elements of Γ). We say that a group G of automorphisms of Δ is an *N-group* if the only inner automorphisms I_a belonging to G are precisely those where $0 \neq a \in \Sigma$.

We may now state the *Fundamental Theorem of the Finite Galois Theory for Division Rings*.

Theorem. *Let Δ be a division ring, G a group of automorphisms of Δ, $\Gamma = G^\dagger$. Suppose that $[\Delta:\Gamma]_L$ is finite. Then the maps † and $*$ define a bijection between the set of all N-subgroups of G and the set of all division subrings of Δ containing Γ.*

The main difference between this theorem and the one for fields is that not all subgroups of the Galois group are allowed, only the so-called *N*-groups. For a field every subgroup of the Galois group is an *N*-group, and we recover the standard version of the Fundamental Theorem.

There is of course much·more to the Galois theory of division rings. The interested reader should consult Jacobson [15].

Galois theory of differential equations

There is a theory, due mainly to Ritt and Kolchin, which does for differential equations what Galois theory does for polynomial equations.

A *differential field* is a field K together with a *derivation* $D:K \to K$ with the properties

$$D(k+l) = D(k)+D(l)$$

$$D(kl) = kD(l)+D(k).l$$

for all $k, l \in K$. The canonical example, and the one of most importance for applications, is where K is the field $\mathbf{C}(x)$ of rational complex functions, and D is the usual derivative. We can define *differential subfields, differential extensions* in the obvious way; the derivations must commute with the inclusion maps. The *field of constants* of K is the set of all $k \in K$ such that $D(k) = 0$; it is a subfield containing the prime subfield. A *differential automorphism* α of K is an automorphism making the diagram

$$
\begin{array}{ccc}
K & \xrightarrow{\;\alpha\;} & K \\
{\scriptstyle D}\Big\downarrow & & \Big\downarrow{\scriptstyle D} \\
K & \xrightarrow[\;\alpha\;]{} & K
\end{array}
$$

commute. If M is a differential extension of K then the *differential Galois group* of $M:K$ is the group of all differential automorphisms of M which leave every element of K fixed. It is usually an infinite group.

So far all we have done is tack the word 'differential' in front of everything. Already we can set up the Galois correspondence: if L is a differential subfield of M we define L^* as a subgroup of the Galois group G of $M:K$; and if H is a subgroup of G we define $H^\dagger$ in the usual way. And $H^\dagger$ is in fact a differential subfield. But the Galois correspondence does not yet work. Call a subgroup H *closed* if $H = H^{\dagger *}$, and a subfield L closed if $L = L^{* \dagger}$. Then $*$ and $\dagger$ define a bijection between closed subgroups and closed subfields. Something important is still lacking: we do not know what the closed fields and subgroups look like.

Now we bring on the differential equations. For $y \in K$ we define

$$y' = D(y), \; y'' = D^2(y), \; \cdots, \; y^{(n)} = D^n(y).$$

Consider the homogeneous linear 'differential equation'

$$y^{(n)} + a_1 y^{(n-1)} + \cdots + a_{n-1} y' + a_n y = 0 \qquad (1)$$

where $a_1, \cdots, a_n \in K$. Suppose that in some larger differential field $u_1, \cdots, u_{n+1}$ are solutions of (1). Then it can be shown that there exist *constants* $c_1, \cdots, c_{n+1} \in K$ such that

$$c_1 u_1 + \cdots + c_{n+1} u_{n+1} = 0$$

where the c_i are not all zero. So there are at most n solutions of (1) linearly independent over constants. A differential extension M of K is a *Picard-Vessiot* extension (for the equation (1)) if:

(1) M is the differential field generated by K together with $u_1, \cdots, u_n$, where the u_i satisfy (1) and are linearly independent over constants,

(2) M has the same field of constants as K.

In the same way that we asked whether the quintic equation could be solved by radicals, we may ask whether a particular differential equation can be solved by specified methods. Two important methods are the following:

(1) *Adjoining an integral.* Adjoin u such that $u' - a = 0$, for $a \in K$.

(2) *Adjoining an exponential of an integral.* Adjoin u such that $u' - au = 0$, for $a \in K$.

The terminology is obvious on considering differential equations for real-valued functions u.

M is a *Liouville extension* of K if there is a chain of subfields

$$K = K_0 \subseteq K_1 \subseteq \cdots \subseteq K_n = M$$

such that each extension $K_{i+1} : K_i$ is either the adjunction of an integral or the exponential of an integral. It can then be proved:

Theorem. *If M is a Liouville extension of K then the differential Galois group is soluble.*

To go further we must introduce more structure on the differential Galois group. It can be considered as an *algebraic group* over the field of constants, that is, a group of matrices (g_{ij}) defined by a system of polynomial equations. Thus the equation $\det(g_{ij}) = 1$ defines the algebraic group of all unimodular matrices. The matrices arise because automorphisms induce linear transformations on the vector space of solutions of the differential equation, spanned by $u_1, \cdots, u_n$.

For a Picard-Vessiot extension it turns out that every intermediate field is closed; while the closed subgroups of the differential Galois group are its *algebraic* subgroups (satisfying additional polynomial equations).

To state our next result we need one more definition. Any algebraic group can be given a topology, called the *Zariski topology*. It need not be connected in this topology, but it certainly splits into connected components. The connected component which contains the identity element of the group is a normal subgroup, called the *component of the identity*.

Theorem. *Suppose that $M:K$ is a Picard-Vessiot extension, such that K has characteristic zero and the field of constants C is algebraically closed. Suppose that M can be embedded in a differential field obtained from K by a finite series of simple algebraic extensions, adjunctions of integrals, or adjunctions of exponentials of integrals. Then the component of the identity of the differential Galois group is soluble.*

Conversely if the differential Galois group of $M:K$ has soluble component of the identity, then M can be obtained from K by a finite normal extension followed by a Liouville extension.

If we consider the differential equation

$$y'' - xy = 0 \tag{2}$$

over the field $\mathbf{C}(x)$ of rational complex functions, it can be shown that the differential Galois group is the full group of unimodular 2×2 matrices over $\mathbf{C}$. This does not have soluble component of the identity, so that (2) cannot be solved by starting with rational functions and performing algebraic operations, integrals, or exponentiation of integrals. This means that (2) cannot be solved by any nice (or even moderately nasty) formula involving standard functions of analysis.

Nevertheless the equation (2), known as *Airy's equation*, has simple power-series solutions:

$$1 + \frac{x^3}{2.3} + \frac{x^6}{2.3.5.6} + \cdots$$

$$x + \frac{x^4}{3.4} + \frac{x^7}{3.4.6.7} + \cdots$$

so that these power series cannot be summed by any closed formula.

The reader who wishes to follow this subject further should consult, in the first instance, Kaplansky [16].

Harder exercises

These exercises are designed for the student to test more deeply his grasp of the topics discussed. Several of them extend the theory beyond the bounds of the text.

1   *The theorem of the primitive element.* If $M:K$ is a finite extension then M is a simple extension $K(\alpha)$ if and only if there are only finitely many intermediate fields.

   (Hint: (1) if K is finite use 16.8. (2) If K is infinite take α to make $[K(\alpha):K]$ maximal. For $\beta \in M \backslash K(\alpha)$ consider the fields $K(\alpha + k\beta)$ as k runs through K, to show β does not exist and $M = K(\alpha)$. (3) If $M = K(\alpha)$ let α have minimum polynomial m over K. If L is an intermediate field let α have minimum polynomial g over L. Show that g determines L uniquely. (4) Show that there are only a finite number of possible g.)

2   Deduce from Exercise 1 that any finite separable extension is a simple extension.

3   Construct a finite extension which is not simple.

4   Let $\gamma = \sqrt{2+\sqrt{2}}$. Show that $\mathbf{Q}(\gamma):\mathbf{Q}$ is normal, with cyclic Galois group. Show that $\mathbf{Q}(\gamma, i) = \mathbf{Q}(\phi)$ where $\phi^4 = i = \sqrt{-1}$.

5 Let θ have minimum polynomial

$$t^3 + at^2 + bt + c$$

over **Q**. Find necessary and sufficient conditions in terms of a, b, c such that $\theta = \phi^2$ where $\phi \in \mathbf{Q}(\theta)$. (Hint: consider the minimum polynomial of ϕ.)

Express $\sqrt[3]{28} - 3$ as a square in $\mathbf{Q}(\sqrt[3]{28})$, and $\sqrt[3]{5} - \sqrt[3]{4}$ as a square in $\mathbf{Q}(\sqrt[3]{5}, \sqrt[3]{2})$. (See Ramanujan [24] p. 329).

6 Let Γ be a finite group of automorphisms of K with fixed field K_0. Let t be transcendental over K. For each $\sigma \in \Gamma$ show there is a unique automorphism σ' of $K(t)$ such that

$$\sigma'(k) = \sigma(k) \qquad (k \in K)$$
$$\sigma'(t) = t.$$

Show that the σ' form a group Γ' isomorphic to Γ, with fixed field $K_0(t)$.

7 Let t be transcendental over K. Show that for any $a, b, c,$ $d \in K$ with $ad - bc \neq 0$ the map $\begin{Bmatrix} a & b \\ c & d \end{Bmatrix}$ defined by

$$t \to (at+b)/(ct+d)$$

determines a K-automorphism of $K(t)$. Show that every K-automorphism arises in this way.

8 Let Γ be the Galois group of $K(t):K$. Show that there is a homomorphism

$$\phi : \mathbf{GL}_2(K) \to \Gamma$$

where $\mathbf{GL}_2(K)$ is the group of all nonsingular 2×2 matrices over K and

$$\phi\left(\begin{pmatrix} a & b \\ c & d \end{pmatrix} \right) = \begin{Bmatrix} a & b \\ c & d \end{Bmatrix}.$$

Show that $\ker(\phi)$ is the set of scalar matrices $\begin{pmatrix} k & 0 \\ 0 & k \end{pmatrix}$ where $0 \neq k \in K$, so that $\Gamma \cong \mathbf{GL}_2(K)/(\text{scalar matrices}) = \mathbf{PGL}_2(K)$, the *projective general linear group*.

9 Let K be a field of characteristic $\neq 2, 3$ containing an element i such that $i^2 = -1$. Consider the automorphisms of $K(t)$ which leave K fixed and map t to any one of $\pm t$, $\pm 1/t$, $\pm i(t+1)/(t-1)$, $\pm i(t-1)/(t+1)$, $\pm(t+i)/(t-i)$, $\pm(t-i)/(t+i)$. Show that these form a group isomorphic to the group of rotations of the regular tetrahedron. Find the fixed field.

(Hint: Consider a sphere of diameter 1 placed with its South Pole at the origin O of the complex plane **C**. Projection from the North Pole defines a map from **C** to this sphere such that the unit circle in **C** becomes the equator. Call the North Pole '∞' and label the other points of the sphere by the corresponding points of **C**. This is the *Riemann Sphere*. Consider the effects of the above maps on this sphere, and in particular on the octahedron whose vertices are 0, ∞, ± 1, $\pm i$. See Klein [41]).

10 Let K have characteristic zero, and let t be transcendental over K. Consider the equations (for x, y, z):

$$x^2 = y + t$$
$$y^2 = z + t \qquad\qquad (*)$$
$$z^2 = x + t.$$

Show that any solution x in an appropriate extension field satisfies either $x^2 = x + t$ or a certain sextic over $K(t)$. Show that the zeros of this sextic can be partitioned into two sets of 3 which are either left invariant or interchanged by any automorphism of the splitting field. Hence solve (*) by radicals. (See Ramanujan [24] p. 327.)

11 Discuss the construction of regular polygons using a ruler, compasses, and an angle trisector. (For example, 9-gons or 13-gons are constructible. Use the trigonometric solution of cubic equations.)

12 If $L:K$ has Galois group $\mathbf{C}_2$ show it is normal. If further K has characteristic $\neq 2$ show that $L = K(\alpha)$ where $\alpha^2 \in K$.

13 Suppose $L:K$ is separable normal of degree 4 with Galois group $\mathbf{C}_2 \times \mathbf{C}_2$, and K has characteristic $\neq 2$. Show that $L = K(\alpha, \beta)$ where $\alpha^2, \beta^2 \in K$.

14 Conversely if K has characteristic $\neq 2$ and $\alpha^2 = a \in K$, $\beta^2 = b \in K$, and none of a, b, ab are squares in K, then $K(\alpha, \beta):K$ has Galois group $\mathbf{C}_2 \times \mathbf{C}_2$.

15 Show that for sufficiently large integers N, and a given prime p, the polynomial

$$x(x - Np^2)(x + Np^2)(x^2 + N^2 p^4) + p$$

cannot be solved by radicals.

16 Let p be an odd prime, ζ a complex pth root of unity. Show that the Galois group of $\mathbf{Q}(\zeta):\mathbf{Q}$ is isomorphic to the multiplicative group $\mathbf{Z}_p \backslash \{0\}$ so is cyclic. Hence show that there is a unique subfield K of $\mathbf{Q}(\zeta)$ such that $[K:\mathbf{Q}] = 2$. Let Γ be the Galois group of $\mathbf{Q}(\zeta):\mathbf{Q}$, and χ the unique homomorphism $\Gamma \to \{\pm 1\}$. Let

$$\alpha = \sum_{s=1}^{p-1} \chi(s) \zeta^s.$$

Show that $\alpha \notin \mathbf{Q}$, $\alpha^2 \in \mathbf{Q}$, and $\alpha^2 = (-1)^{(p-1)/2} p$. Hence show that $K = \mathbf{Q}(\alpha)$.

17 Suppose f is a separable irreducible quartic polynomial over K, where K has characteristic zero; and let α be a zero of f. Show that there are no fields strictly between $K(\alpha)$ and K if and only if the Galois group of f is $\mathbf{A}_4$ or $\mathbf{S}_4$.

18 If $t^3 + at + b$ is irreducible over a finite field K, show that $-4a^3 - 27b^2$ is a square in K.

19 Prove that in a finite field every element is a sum of two squares.

20 Let Γ be the group of order 8 given by generators and relations

$$\langle \alpha, \beta, \gamma : \alpha^2 = \beta^2 = \gamma^2 = 1, \beta\gamma = \alpha\gamma\beta, \alpha\beta = \beta\alpha, \alpha\gamma = \gamma\alpha \rangle.$$

Let K be a field of characteristic $\neq 2$ and let $L = K(r, s)$ where $r^2 = a \in K$, $s^2 = b \in K$, and $[L:K] = 4$. Show that the following statements are equivalent.

(1) There is a normal extension $M:K$ with Galois group Γ such that L is the fixed field of $\{1, \alpha\}$.

(2) There is a solution of the equation

$$ax^2 + by^2 = z^2$$

where $x, y, z \in K$, $xyz \neq 0$.

(Hint: If $\beta(s) = s$ write the fixed field of $\{1, \beta\}$ in the form $K(s, t)$ where $t^2 = c \in K(s)$. Then look at $\gamma(t)/t$ and $t \cdot \gamma(t)$.)

Selected
solutions

In this section we give solutions or hints for a selection of the exercises.

1.2 No. $2\mathbf{Z}$ has no multiplicative identity.

1.3 $\mathbf{Z}_7$ is a field and an integral domain. $\mathbf{Z}_6$ and $\mathbf{Z}_8$ are neither.

1.4 Addition and multiplication are commutative.

1.6 It is not isomorphic to $\mathbf{Z}_4$ since $\mathbf{Z}_4$ is not a field. There is just 1 field with 4 elements (up to isomorphism).

1.9 (a) $q = t^4 - 7t - 1$, $r = 49t + 12$.
(b) $q = 1$, $r = 1$.
(c) $q = 2t^2 - \frac{27}{2}t + \frac{137}{4}$, $r = -\frac{697}{4}$.
(d) $q = t^2 - 1$, $r = 0$.
(e) $q = 4t^4 - 2t^3 + 4t + 2$, $r = -2t + 2$

1.10 (a) 1. (b) 1. (c) 1. (d) $t + 2$. (e) $t - 1$.

1.13 $\mathbf{C}_4, \mathbf{C}_2, \mathbf{C}_2 \times \mathbf{C}_2, \mathbf{C}_2 \times \mathbf{C}_2 \times \mathbf{C}_2$.

1.14 2, 3, 4, 6, 8, 12, 24 and only these.

1.15 ?FFFTTFTFF.

2.1 The irreducible ones are (b), (c), (d), (f).

2.2 (a) $(t^2 + \sqrt{2}t + 1)(t^2 - \sqrt{2}t + 1)$. (e) $(t - 1)(t^2 - 6t - 3)$.
(g) $(t - 3)(t^2 + 3t + 9)$. (h) $(t + 1)(t + \beta)$.

2.3 Use Corollary 2.8.

2.6 $a^2 - 4b$ is a square if and only if the polynomial is reducible.

2.7 As for 2.6.

2.8 They are not isomorphic since $\mathbf{Z}_p$ is finite and $\mathbf{Z}_p(t)$ is infinite.

2.9 (a) $s_1^2 - 2s_2$. (b) $s_1^3 - 3s_1 s_2 + 3s_3$. (c) $s_1 s_2 - 3s_3$. (d) s_3^2.
(e) $2s_1^2 - 4s_2$. (f) $(4(s_1^2 - 3s_2)^3 - (4s_1^3 - 9s_2 s_1 + 27s_3)^2)/27$
(g) The polynomial is not symmetric.

2.10 You get back the original cubic equation.

2.11 FFTTFTTFFF.

3.2 (a) $\mathbf{Q}$. (b) $\mathbf{Q}$. (c) $\{p + qi : p, q \in \mathbf{Q}\}$.
(d) $\{p + q\sqrt{2} + ri + si\sqrt{2} : p, q, r, s \in \mathbf{Q}\}$.
(e) $\{p + q\sqrt{2} + r\sqrt{3} + s\sqrt{6} : p, q, r, s \in \mathbf{Q}\}$.
(f) $\mathbf{R}$. (g) $\mathbf{C}$.

3.3 (a) $\{p + q\sqrt{2} : p, q \in \mathbf{Q}\}$. (b) $\{p + qi : p, q \in \mathbf{Q}\}$.
(c) $\{p + q\alpha + r\alpha^2 : p, q, r \in \mathbf{Q}\}$.
(d) $\{p + q\sqrt{5} + r\sqrt{7} + s\sqrt{35} : p, q, r, s \in \mathbf{Q}\}$.
(e) $\{p + qi\sqrt{11} : p, q \in \mathbf{Q}\}$.

3.4 (a) Expressions with no odd powers of t. (b) $K(t)$.
(c) Expressions with only t^n where n is a multiple of 5.
(d) Same as (a).

3.5 Algebraic: those in 3.3. Transcendental: those in 3.4. All are simple, although 3.3(d) is not obviously so.

3.8 (a) $t^2 + 1$. (b) $t^2 + 1$. (c) $t^2 - 2$. (d) $t^2 - t - 1$. (e) $t^2 + t + 1$.
(f) $t^2 + t + 1$. (g) $u^2 - t - 1$ (considered as a polynomial in u).

3.10 Only in case (b).

3.18 Yes.

3.19 No.

3.20 TFFFFTTFFF.

4.1 (a) ∞. (b) ∞. (c) 1. (d) 3. (e) 8. (f) 2. (g) 7.

4.7 Under the requirement of continuity no such functions exist.

4.11 Whenever the element of L is non-zero.

4.13 Yes.

4.16 The degree is 4.

4.18 FTFTTFFTTF.

5.4 Yes.

5.10 TTTFTTTFFT.

6.9 TFTTTFFTFT.

7.1   (a) $\sqrt{2} \to \pm\sqrt{2}$.   (b) $\alpha \to \alpha$.
       (c) $\sqrt{2} \to \pm\sqrt{2}$, $\sqrt{3} \to \pm\sqrt{3}$ with any choice of signs.

7.2 $\mathbf{C}_2$, $\mathbf{C}_1$, $\mathbf{C}_2 \times \mathbf{C}_2$.

7.3   (a) and (c).

7.4   The group is $\mathbf{C}_2$ and the Galois correspondence is a bijection.

7.6 $\mathbf{Q}(\sqrt{2},\sqrt{3},\sqrt{5})$, $\mathbf{Q}(\sqrt{2},\sqrt{3})$, $\mathbf{Q}(\sqrt{2},\sqrt{5})$, $\mathbf{Q}(\sqrt{3},\sqrt{5})$, $\mathbf{Q}(\sqrt{2})$, $\mathbf{Q}(\sqrt{3})$, $\mathbf{Q}(\sqrt{5})$, $\mathbf{Q}$. The group is $\mathbf{C}_2 \times \mathbf{C}_2 \times \mathbf{C}_2$ and the correspondence is a bijection.

7.7   T T F T F F F T F T.

8.1 $\mathbf{Q}(e^{2\pi i/3})$, $\mathbf{Q}(i\sqrt{2},i\sqrt{3})$, $\mathbf{Q}(\sqrt{2}, e^{\pi i/3})$.

8.2   2, 4, 12.

8.5   All the fields have 25 elements and are isomorphic.

8.7   Take care to resolve into irreducibles before applying proposition 8.6 in characteristic $p > 0$.

8.8   (b), (d), and (e).

8.9   False for any degree > 2. (Extend $\mathbf{Q}$ by a real nth root of 2.)

8.12   T T T F F T T T F F.

9.4   There is only 1.

9.5   There is only 1.

9.6   It fails if K is not a field.

9.7   T F T F F T F T T.

10.1   False for $K(t)$, with $t \to t^2$.

10.2   (a) $\mathbf{Q}(\alpha, e^{2\pi i/5})$.   (b) $\mathbf{Q}(\beta, e^{2\pi i/7})$.   (c) $\mathbf{Q}(\sqrt{2}, \sqrt{3})$. (d) $\mathbf{Q}(\alpha, \sqrt{2}, e^{2\pi i/3})$. (e) splitting field for $t^3 - 3t^2 + 3$.

10.3   (a) $\mathbf{C}_1$.   (b) $\mathbf{C}_1$.   (c) $\mathbf{C}_2 \times \mathbf{C}_2$.   (d) $\mathbf{C}_2$.

10.4   (a) $\langle \gamma, \delta : \gamma^5 = \delta^4 = 1, \delta^{-1}\gamma\delta = \gamma^3 \rangle$.
       (b) $\langle \gamma, \delta : \gamma^7 = \delta^6 = 1, \delta^{-1}\gamma\delta = \gamma^3 \rangle$.
       (c) $\mathbf{C}_2 \times \mathbf{C}_2$.
       (d) $\mathbf{C}_2 \times \mathbf{S}_3$.

10.6   8.

10.7   F T T F F T F T T F.

12.1   (a) $\mathbf{C}_2 \times \mathbf{C}_2$.   (b) $\mathbf{C}_2$.   (c) $\mathbf{C}_2 \times \mathbf{C}_2$.

13.1 $\{1, a, \cdots, a^{n-1}\}$ is cyclic and normal with $\mathbf{C}_2$ quotient.

13.2 $\mathbf{S}_n$ has a subgroup isomorphic to $\mathbf{A}_5$.

13.6   Use Cauchy's theorem 13.14.

13.14 FTTTFFFTTT.

14.3 C_{p-1}.

14.4 Mimic the proof of Theorem 14.8.

14.6 Use proposition 10.2.

14.9 TTFTTTTTT.

15.2 (a) 4. (b) 2. (c) 0.

15.11 TFTTFFFFFT.

16.1 2, 3, 4, 5, 17, 65536, 65537, 83521, 103823, $2^{19937} - 1$.
The last is at the time of writing the largest known
prime number.

16.11 $n = 1$ or 2 only.

16.12 FTTTTTTFFF.

17.4 641 divides $5^4 \cdot 2^{28} + 2^{32}$ and $5^4 \cdot 2^{28} - 1$. Their difference is F_5 (proof due to G. T. Bennett).

17.6 1, 2, 3, 4, 5, 6, 8, 10, 12, 15, 16, 17, 20, 24, 30, 32, 34, 40,
48, 51, 60, 64, 68, 80, 96.

17.8 TFTTFTTTTF.

18.4 $-1 = i^2$ is a square, so positive: contradiction. If we
change the operations we can give **C** an ordered field
structure: put **C** in 1–1 correspondence with **R** and
use the operations on **R** to induce operations on **C**.

18.10 TTFTTTFTT.

References

Galois Theory

1. Adamson, I. T. (1964), *Introduction to Field Theory*, Oliver and Boyd, Edinburgh.
2. Artin, E. (1948), *Galois Theory*, Notre Dame.
3. Jacobson, N. (1964), *Lectures in Abstract Algebra* vol. 3 – *Theory of Fields and Galois Theory*, Van Nostrand, Princeton.
4. Kaplansky, I. (1969), *Fields and Rings*, Chicago.
5. Van der Waerden, B. L. (1953), *Modern Algebra* (2 vols.), Ungar, New York.

Additional Mathematical Material

6. Carter, R. W. (1965), 'Simple groups and simple Lie algebras', *J. London Math. Soc.*, **40**, 193–240.
7. Chase, S. U., Harrison, D. K., and Rosenberg, A. (1965), 'Galois Theory and Cohomology of Commutative Rings', *Memoirs of the Amer. Math. Soc.*, **52**, Providence R.I.
8. Cohn, P. M. (1961), 'Quadratic extension of Skew Fields', *Proc. London Math. Soc.*, (3) **11**, 531–556.
9. Cundy, H. M. and Rollett, A. P. (1961), *Mathematical Models*, Oxford.

10. Gödel, K. (1962), *On formally undecidable propositions of Principia Mathematica and related systems*, Oliver and Boyd, Edinburgh.
11. Halmos, P. R. (1958), *Finite-dimensional Vector Spaces*, Van Nostrand, Princeton.
12. Hardy, G. H. (1960), *A course of Pure Mathematics*, Cambridge.
13. Hardy, G. H. and Wright, E. M. (1962), *The Theory of Numbers*, Oxford.
14. Hartley, B. and Hawkes, T. O. (1970), *Rings, Modules and Linear Algebra*, Chapman and Hall, London.
15. Jacobson, N. (1964), *Structure of Rings*, Amer. Math. Soc. Colloquium Publications XXXVII, Providence R.I.
16. Kaplansky, I. (1957), *An introduction to Differential Algebra*, Hermann, Paris.
17. Kobelev, V. V. (1970), 'Proof of Fermat's Last Theorem for all prime exponents less than 5500', *Dokl. Akad. Nauk SSSR* **190**, 767–768 (Russian); translated in *Soviet Math.*, **11** (1970) 188–189.
18. laChyl, Eve (1968), 'Simple Groups', *Manifold* **2**.
19. Ledermann, W. (1961), *The Theory of Finite Groups*, Oliver and Boyd, Edinburgh.
20. MacDonald, I. D. (1968), *The Theory of Groups*, Oxford.
21. Mendelson, E. (1964), *Introduction to Mathematical Logic*, Van Nostrand, Princeton.
22. Mordell, L. J. (1969), *Diophantine Equations*, Academic Press, London and New York.
23. Oldroyd, J. C. (1955), *Eureka* **18**, 20.
24. Ramanujan, S. (1962), *Collected Papers of Srinivasa Ramanujan*, Chelsea, New York.
25. Salmon, G. (1885), *Lessons introductory to the modern Higher Algebra*, Hodges, Figgis, and Co., Dublin.
26. Titchmarsh, E. C. (1960), *The Theory of Functions*, Oxford.
27. Tits, J. (1970), *Groupes Finis Simples Sporadiques*, Séminaire Bourbaki 22e année 1969/70 no 375.

Historical Material

28. Bell, E. T. (1965), *Men of Mathematics* (2 vols.), Penguin, Harmondsworth, Middlesex.
29. Bertrand, J. (1899), 'La vie d'Évariste Galois, par P. Dupuy', *Bull. des sciences mathématiques*, 198–212.
30. Bortolotti, E. (1925), 'L'algebra nella scuola matematica bolognese del secolo XVI', *Periodico di Matematica* (4) **5**, 147–184.
31. Bourgne, R. and Azra, J.-P. (1962), *Écrits et mémoires mathématiques d'Évariste Galois*, Gauthier-Villars, Paris.
32. Bourbaki, N. (1969), *Éléments d'Histoire des Mathématiques*, Hermann, Paris.
33. Cardano, G. (1931), *The Book of my Life*, Dent, London.
34. Clifford, W. K. (1968), *Mathematical papers*, Chelsea, New York.
35. Coolidge, J. L. (1963), *The Mathematics of great Amateurs*, Dover, New York.
36. Dalmas, A. (1956), *Évariste Galois révolutionnaire et géomètre,* Fasquelle, Paris.
37. Dupuy, P. (1896), 'La vie d'Évariste Galois', *Annales de l'École Normale* (3) **13**, 197–266.
38. Galois, E. (1897), *Oeuvres mathematiques d'Évariste Galois*, Gauthier-Villars, Paris.
39. Gauss, C. F. (1966), *Disquisitiones Arithmeticae*, Yale University Press, New Haven.
40. Huntingdon, E. V. (1905), *Trans. Amer. Math. Soc.* **6**, 181.
41. Klein, F. (1913), *Lectures on the Icosahedron and the Solution of Equations of the fifth degree*, Kegan Paul, London.
42. Klein, F. *et al.* (1962), *Famous Problems and other monographs*, Chelsea, New York.
43. Kollros, L. (1949), *Évariste Galois*, Birkhäuser, Basel.
44. Midonick, H. (1965), *The Treasury of Mathematics* (2 vols.), Penguin, Harmondsworth, Middlesex.

45. Mordell, L. J. *Three lectures on Fermat's Last Theorem*, printed in the same volume as reference [43].

46. Richelot, F. J. (1832), 'De resolutione algebraica aequationis $x^{257} = 1$, sive de divisione circuli per bisectionam anguli septies repetitam in partes 257 inter se aequales commentatio coronata', *Crelle's Journal* IX, 1–26, 146–161, 209–230, 337–356.

47. Richmond, H. W. (1893), *Quart. J. Math.*, **26**, 206–207 and *Math. Ann.*, **67** (1909) 459–461.

48. Struik, D. J. (1962), *A concise history of Mathematics*, Bell, London.

49. Taton, R. (1947), 'Les relations d'Évariste Galois avec les mathématiciens de son temps', Cercle International de synthèse, *Revue d'histoire des sciences et de leurs applications*, **1**, 114.

Index